Characterization and Modelling of the Deformation and Failure of Engineering Metallic Materials

Characterization and Modelling of the Deformation and Failure of Engineering Metallic Materials

Guest Editors

Hui Wang
Lihong Su
Ebad Bagherpour
Qiang Xing

MDPI • Basel • Beijing • Wuhan • Barcelona • Belgrade • Novi Sad • Cluj • Manchester

Guest Editors
Hui Wang
School of Mechanical Engineering
Nantong University
Nantong
China

Lihong Su
School of Mechanical, Materials, Mechatronic and Biomedical Engineering
University of Wollongong
Wollongong
Australia

Ebad Bagherpour
Brunel Centre for Advanced Solidification Technology
Brunel University London
Uxbridge
UK

Qiang Xing
School of Mechanical Engineering
Nantong University
Nantong
China

Editorial Office
MDPI AG
Grosspeteranlage 5
4052 Basel, Switzerland

This is a reprint of the Special Issue, published open access by the journal *Crystals* (ISSN 2073-4352), freely accessible at: https://www.mdpi.com/journal/crystals/special_issues/1A1HP10Z87.

For citation purposes, cite each article independently as indicated on the article page online and as indicated below:

Lastname, A.A.; Lastname, B.B. Article Title. *Journal Name* **Year**, *Volume Number*, Page Range.

ISBN 978-3-7258-5711-1 (Hbk)
ISBN 978-3-7258-5712-8 (PDF)
https://doi.org/10.3390/books978-3-7258-5712-8

Contents

About the Editors

Hui Wang

Hui Wang received his Ph.D. in mechanical engineering in 2016 from the University of Wollongong, Australia. After working there as an associate research fellow for two years, he joined the Institute of Industrial Science (IIS) at the University of Tokyo, Japan, as a project research fellow. In 2023, he joined the School of Mechanical Engineering, Nantong University, as a Professor. His research focuses on metal deformation and failure, metal forming, and the microstructure–property relationship. His research methods include experimental techniques, numerical modelling, and machine learning. He has strong research collaborations with researchers from Japan, South Korea, Australia, and China. He has published around 50 high-quality journal papers, and he has often presented his research at international conferences and industrial workshops as an invited speaker.

Lihong Su

Lihong Su received her Ph.D. in Mechanical Engineering in 2013 from the University of Wollongong, Australia, where she is currently working as an Australian Research Council (ARC)-funded Discovery Early Career Researcher Award (DECRA) research fellow and lecturer. She leads a research team on the 'Development and application of novel structural and functional materials with advanced manufacturing techniques'. Her major research interests include additive manufacturing, advanced rolling, severe plastic deformation, nano/ultrafine-grained materials, 2D materials, XRD and neutron diffraction, texture and residual stress analysis, advanced modelling, and characterisations. She has published around 100 high-quality journal and conference papers, including contributions of 11 invited or keynote talks in international conferences. She has been awarded 13 competitive research grants with USD 6.8 million of funding and 5 prestigious research awards, including the ARC DECRA Fellowship, the Japan Society for the Promotion of Science (JSPS) Fellowship, the France–Australia Science Innovation Collaboration (FASIC) Program Early Career Fellowship, and the Australia Defence Materials Technology Centre (DMTC) Project Leadership Award.

Ebad Bagherpour

Ebad Bagherpour is a Lecturer at the Brunel Centre for Advanced Solidification Technology (BCAST), Brunel University London. His research focuses on deformation mechanisms, advanced forming technologies, the cyclic response of metals, microstructure–property relationships, and thermomechanical processing of metallic materials. He also works extensively on sustainability and circularity in the metals sector, with particular interest in the recovery, rejuvenation, and extended service life of metallic components. With over 15 years of combined academic and industrial experience, Ebad Bagherpour has contributed extensively to the field of materials science through his investigations into the mechanical behaviour of metals and the development of innovative processing techniques. He has published widely in peer-reviewed journals and actively contributes to international research collaborations. Ebad Bagherpour earned his PhD in Materials Science and Engineering through a joint programme between Shiraz University and Semnan University, with part of his doctoral research carried out at Doshisha University, Japan. He also holds MSc and BSc degrees in the same field from Shiraz University. He is a Fellow of the Institute of Materials, Minerals and Mining (FIMMM), an Associate Editor for *Discover Applied Science* (Springer Nature), and a member of the Elsevier Advisory Panel. His editorial roles include serving as a Guest Editor for *Crystals* (MDPI), and an Editor for *Modern Materials Science and Technology* and *Engineering and Solid Mechanics*. Dr.

Bagherpour remains committed to advancing the field of materials science through research, scholarly publishing, and active engagement with the scientific community.

Qiang Xing

Qiang Xing is currently an Associate Professor in the School of Mechanical Engineering at Nantong University. He received his Ph.D. from Nanjing University of Aeronautics and Astronautics, China, in 2014, and subsequently joined Nantong University. His research interests include micro/nano-manufacturing, visual perception, and bionics, as well as complex intelligent mechatronic systems. He has published nearly 30 academic papers and serves as a reviewer for several scholarly journals.

Preface

Metallic materials remain the backbone of structural applications across industries. Their reliability and performance are intimately linked to their microstructural characteristics, governing deformation mechanisms and eventual failure. Understanding these complex phenomena requires both precise experimental characterization and robust modelling frameworks that span multiple length and time scales.

This Reprint, "Characterization and Modelling of the Deformation and Failure of Engineering Metallic Materials," aims to provide a comprehensive overview of recent progress in this field. It covers state-of-the-art experimental techniques, from electron microscopy and digital image correlation to atom probe tomography, that enable insights into the microstructural origins of mechanical behavior. In parallel, it examines numerical modelling approaches, including finite element methods, crystal plasticity, dislocation dynamics, and atomistic simulations, which together form a multiscale perspective on metallic deformation and failure.

The motivation for this work stems from the growing need to bridge the gap between advanced experimental observations and predictive computational models, ultimately enabling the design of safer, stronger, and more sustainable materials. The Reprint is addressed to graduate students, researchers, and engineers in materials science, mechanical engineering, and related disciplines, offering both fundamental knowledge and practical guidance for future innovations.

Hui Wang, Lihong Su, Ebad Bagherpour, and Qiang Xing
Guest Editors

crystals

Editorial

Characterization and Modelling of the Deformation and Failure of Engineering Metallic Materials

Hui Wang [1,2,*], Lihong Su [3,*], Ebad Bagherpour [4,*] and Qiang Xing [1]

1 School of Mechanical Engineering, Nantong University, Nantong 226019, China
2 Institute of Industrial Science (IIS), The University of Tokyo, Kashiwa 277-8574, Chiba, Japan
3 School of Mechanical, Materials, Mechatronic and Biomedical Engineering, University of Wollongong, Wollongong, NSW 2522, Australia
4 Brunel Centre for Advanced Solidification Technology (BCAST), Brunel University London, Uxbridge UB8 3PH, UK
* Correspondence: hw737ntu@163.com (H.W.); lihongsu@uow.edu.au (L.S.); ebad.bagherpour@brunel.ac.uk (E.B.)

Metallic materials are at the heart of modern industry and infrastructure, valued for their outstanding strength, ductility, and other excellent mechanical properties [1,2]. The deformation and failure in these materials are inherently multiscale and multifactorial phenomena [2,3]. At the atom- or micro-scale, they originate from dislocation motion, twinning, phase transformation, void nucleation, etc. [1,4]. Plastic deformation and failure are strongly shaped by how different microstructural features interact with one another [4,5]. As we push for lighter structures [6–8], greater durability in harsh environments [9,10], and longer service lives [11–13], understanding how metals deform and fail has become more important—and more challenging—than ever. For this purpose, both advanced characterization techniques and numerical modelling methods have been developed.

Material characterization techniques have advanced quickly in recent decades [14,15], which offers insights into deformation mechanisms at a wide range of length and time scales. At the micro-scale, in situ mechanical testing using electron microscopy allows for the direct observation of dislocation activity [14], phase evolution [16], and crack propagation during loading [15]. Techniques like Scanning electron microscopy (SEM), electron backscatter diffraction (EBSD), and transmission electron microscopy (TEM) are widely used to reveal local crystal orientations, microstructural features, and areas of strain, offering a detailed understanding of how materials deform at the grain level [15]. At larger scales, digital image correlation (DIC) techniques offer full-field strain mapping during mechanical tests [17,18], revealing heterogeneity in strain distribution and early signs of localized deformation. Meanwhile, X-ray diffraction, particularly using synchrotron radiation, has enabled the measurement of internal stresses, lattice strains, and phase distributions [19]. Emerging techniques, such as 3D tomography, atom probe tomography (APT), and a focused ion beam (FIB), allow for the three-dimensional reconstruction of microstructures [19–22].

Numerical modelling techniques have also progressed significantly. The finite element method (FEM) remains a cornerstone of mechanical simulation. The conventional FEM relies on phenomenological constitutive models, which does not consider microstructural factors [23–25]. When crystal plasticity models are integrated into finite element simulations, crystal plasticity finite element methods (CPFEMs) can account for the orientation of crystals and the specific slip systems that operate during deformation [26–28]. The

 https://doi.org/10.3390/cryst15070642

microstructure in a CPFEM can originate from EBSD data, ensuring more accurate predictions [29,30]. At a lower scale, discrete dislocation dynamics (DDD) simulations capture the collective behaviour of dislocations and their interactions [31], while molecular dynamics (MD) simulations provide atomistic insights into mechanisms such as dislocation nucleation, phase transformation, and grain boundary migration [32,33]. Though computationally intensive, these numerical approaches are invaluable for exploring deformation phenomena that are difficult to observe experimentally. This Special Issue has collected works regarding the recent progress concerning on the above-mentioned topics.

Zhang et al. [34] used cavitation water jet peening (CWJP) to strengthen the surface of the 7075 aluminum alloy. The process increased surface hardness from 109.2 Hv to 144.0 Hv, with the strengthening effect extending up to 600 μm deep. In another study by Ding et al. [35], laser shock peening was also used to improve the fatigue life of 1Cr18Ni9Ti. It was found that the 3J category produced the best result due to the optimal grain refinement and residual compressive stress. Abdi et al. [36] applied twin parallel channel angular extrusion (TPCAE) to an AZ91 cast magnesium alloy to improve its hydrogen storage properties. The TPCAE process was conducted at temperatures from 340 °C to 200 °C, and the hydrogen absorption and desorption tests were measured at 250 °C, 300 °C, and 350 °C. Three TPCAE passes at 250 °C resulted in the best absorption capacity, i.e., 6.1 wt.% within a time span of 2000s. To enhance the welding quality of aluminum and steel, Zhang et al. [37] placed a copper and a nickel coating on the surface of the former and latter, respectively. This interlayer copper–nickel binary coating suppressed the formation of brittle intermetallic compounds, and the welding quality was increased by 56% compared to the uncoated aluminum–steel welding.

Chen et al. [38] conducted dry sliding tribometric tests with different loads to evaluate the mild–severe wear transition of the 2095 Al-Li alloy. The wear rate grew slowly when the load was 2–4N, grew fast at a medium load (8–16N), and grew gradually at loads from 32N to 40N. SEM mapping shows that the abrasion and oxidation were significant during the transition from mild to severe wear, and the tribo-induced plastic deformation of the substrate is the reason for this wear transition. Wang et al. [39] used multiple misorientation parameters, derived from EBSD mapping, to evaluate the plastic damage, including the grain reference orientation deviation (GROD), grain orientation spread (GOS), grain orientation spread over the grain diameter (GOS/D), and the geometrically necessary dislocation (GND). It was found that the GOS/D was a reasonable indicator for the plastic damage in 316 steels. The underlying mechanism for plastic deformation and damage is dislocation movement, and thus understanding the behaviours of dislocation movement is critically important. Chang et al. [40] used an indentation stress relaxation process to study the dislocation velocity and stress exponent in commercial pure aluminum, and this exponent was found to be 2.5 ± 0.5 for ambient temperature.

Shen et al. [30] conducted CPFEM simulations to investigate the influence of ultrafine-grained (UFG) austenite on the mechanical properties of medium-Mn steel by comparing the stress and strain between the samples with and without austenite. The 3D EBSD-scanned microstructure was modelled to ensure high fidelity, where the grain morphology, crystallographic orientation, and phase composition were preserved. The UFG austenite was found to be a main contributor to the ductility and strength of medium-Mn steel. CPFEM simulations were carried out to evaluate the sensitivity of the nanoindentation stress relaxation methodology to the dislocation velocity-stress exponent [40]. Compared to the uniform-field and mean-field methods, the full-field CPFEM requires a large amount of computation time, since homogenization is not theoretically assumed. Liu et al. [41] utilized a Submodel method to exceedingly enhance the mesh resolution in areas of interest, and the CPFEM simulations successfully captured the weak texture components and local deformation. Qin et al. [33] performed molecular dynamics simulations to systemically study the behaviours of crack propagation in the Ni-based superalloy. It was found that high-stress triaxiality promoted brittle cracking, while low-stress triaxiality enhanced the plastic crack; the cracks grow slowly when they are located in the γ and γ' phase; the γ/γ' phase interface hinders the crack propagation when the crack is perpendicular to the γ/γ' interface.

As metal processing techniques continue to evolve, we are seeing the development of more advanced materials with increasingly complex microstructures. These advances call for equally sophisticated tools to understand how structure influences performance. This Special Issue aims to present state-of-the-art developments in this vibrant field, and to deepen our understanding of how metals deform and fail. As Guest Editors, we hope you find this Special Issue helpful and inspiring. Finally, we are deeply grateful to the authors for their publications, to the reviewers for their time and comments, and to the Editorial team for their suggestions and support.

Author Contributions: The authors equally contributed to this work. All authors have read and agreed to the published version of the manuscript.

Funding: L.S. acknowledges the financial support from the Australian Research Council (ARC) through the Discovery Early Career Researcher Award (DECRA) fellowship (No. DE180100124). H.W. was supported by the Large Instruments Open Foundation of Nantong University (KFJN2428), the Test Centre of Nantong University.

Conflicts of Interest: The authors declare no conflicts of interest.

References

1. Singh, M.; Kumar, P.; Biswas, A. An overview of recent developments in Al metal matrix nanocomposites for strength-ductility synergy. *Mater. Today Proc.* **2022**, *80*, 168–175. [CrossRef]
2. Sohrabi, M.J.; Kalhor, A.; Mirzadeh, H.; Rodak, K.; Kim, H.S. Tailoring the strengthening mechanisms of high-entropy alloys toward excellent strength-ductility synergy by metalloid silicon alloying: A review. *Prog. Mater. Sci.* **2024**, *144*, 101295. [CrossRef]
3. Zhao, L.; Zheng, W.; Hu, Y.; Guo, Q.; Zhang, D. Heterostructured metal matrix composites for structural applications: A review. *J. Mater. Sci.* **2024**, *59*, 9768–9801. [CrossRef]
4. Wang, S.; Hu, Z.; Huang, Z.; Gao, B.; Chen, X.; Hu, J.; Zhu, Y.; Li, Y.; Zhou, H. New deformation mechanism and strength-ductility synergy in pure titanium with high density twin. *Int. J. Plast.* **2024**, *174*, 103908. [CrossRef]
5. Gupta, A.; Khatirkar, R.; Singh, J. A review of microstructure and texture evolution during plastic deformation and heat treatment of β-Ti alloys. *J. Alloy Compd.* **2022**, *899*, 163242. [CrossRef]
6. Georgantzia, E.; Gkantou, M.; Kamaris, G.S. Aluminium alloys as structural material: A review of research. *Eng. Struct.* **2021**, *227*, 111372. [CrossRef]
7. Verma, R.P.; Lila, M.K. A short review on aluminium alloys and welding in structural applications. *Mater. Today Proc.* **2021**, *46*, 10687–10691. [CrossRef]
8. Bai, J.; Yang, Y.; Wen, C.; Chen, J.; Zhou, G.; Jiang, B.; Peng, X.; Pan, F. Applications of magnesium alloys for aerospace: A review. *J. Magnes. Alloy* **2023**, *11*, 3609–3619. [CrossRef]
9. Ou, Y.; Wang, H.; Zhao, Y.; Zhou, Q.; Luo, C.; Hua, Q.; Ouyang, X.; Zhang, S. Recent advances and strategies for high-performance coatings. *Prog. Mater. Sci.* **2023**, *136*, 101125. [CrossRef]
10. Zhang, J.; Miao, J.; Balasubramani, N.; Cho, D.H.; Avey, T.; Chang, C.-Y.; Luo, A.A. Magnesium research and applications: Past, present and future. *J. Magnes. Alloy* **2023**, *11*, 3867–3895. [CrossRef]
11. Liu, H.; Yu, H.; Guo, C.; Chen, X.; Zhong, S.; Zhou, L.; Osman, A.; Lu, J. Review on Fatigue of Additive Manufactured Metallic Alloys: Microstructure, Performance, Enhancement, and Assessment Methods. *Adv. Mater.* **2023**, *36*, e2306570. [CrossRef] [PubMed]
12. Zheng, Z.; Zhan, M.; Fu, M.W. Microstructural and geometrical size effects on the fatigue of metallic materials. *Int. J. Mech. Sci.* **2022**, *218*, 107058. [CrossRef]
13. Farh, H.M.H.; Seghier, M.E.A.B.; Zayed, T. A comprehensive review of corrosion protection and control techniques for metallic pipelines. *Eng. Fail. Anal.* **2022**, *143*, 106885. [CrossRef]

14. Li, S.; Powell, C.A.; Mathaudhu, S.; Gwalani, B.; Devaraj, A.; Wang, C. Review of recent progress on in situ TEM shear deformation: A retrospective and perspective view. *J. Mater. Sci.* **2022**, *57*, 12177–12201. [CrossRef]
15. Gussev, M.; McClintock, D.; Byun, T.; Lach, T. Recent progress in analysis of strain-induced phenomena in irradiated metallic materials and advanced alloys using SEM-EBSD in-situ tensile testing. *Curr. Opin. Solid State Mater. Sci.* **2023**, *28*, 101132. [CrossRef]
16. Zhao, H.; Zhu, Y.; Ye, H.; He, Y.; Li, H.; Sun, Y.; Yang, F.; Wang, R. Atomic-Scale Structure Dynamics of Nanocrystals Revealed by In Situ and Environmental Transmission Electron Microscopy. *Adv. Mater.* **2023**, *35*, e2206911. [CrossRef]
17. Weidner, A.; Biermann, H. Review on Strain Localization Phenomena Studied by High-Resolution Digital Image Correlation. *Adv. Eng. Mater.* **2021**, *23*, 2001409. [CrossRef]
18. Janeliukstis, R.; Chen, X. Review of digital image correlation application to large-scale composite structure testing. *Compos. Struct.* **2021**, *271*, 114143. [CrossRef]
19. Guo, E.; Du, Z.; Chen, X.; Chen, Z.; Kang, H.; Cao, Z.; Lu, Y.; Wang, T. Development of magnesium alloys: Advanced characterization using synchrotron radiation techniques. *J. Mater. Sci. Technol.* **2024**, *195*, 93–110. [CrossRef]
20. Li, P.; Chen, S.; Dai, H.; Yang, Z.; Chen, Z.; Wang, Y.; Chen, Y.; Peng, W.; Shan, W.; Duan, H. Recent advances in focused ion beam nanofabrication for nanostructures and devices: Fundamentals and applications. *Nanoscale* **2020**, *13*, 1529–1565. [CrossRef]
21. Hu, R.; Jin, S.; Sha, G. Application of atom probe tomography in understanding high entropy alloys: 3D local chemical compositions in atomic scale analysis. *Prog. Mater. Sci.* **2022**, *123*, 100854. [CrossRef]
22. Baba, N.; Hata, S.; Saito, H.; Kaneko, K. Three-dimensional electron tomography and recent expansion of its applications in materials science. *Microscopy* **2022**, *72*, 111–134. [CrossRef]
23. Wang, H.; Su, L.; Yu, H.; Lu, C.; Tieu, A.K.; Liu, Y.; Zhang, J. A new finite element model for multi-cycle accumulative roll-bonding process and experiment verification. *Mater. Sci. Eng. A* **2018**, *726*, 93–101. [CrossRef]
24. Zhang, Z.; Gu, G.X. Finite-Element-Based Deep-Learning Model for Deformation Behavior of Digital Materials. *Adv. Theory Simulations* **2020**, *3*, 2000031. [CrossRef]
25. Bate, P.; Ashby, M.F.; Humphreys, F.; Sellar, C.; Shercliff, H.; Stowell, M. Modelling deformation microstructure with the crystal plasticity finite–element method. *Philos. Trans. R. Soc. A Math. Phys. Eng. Sci.* **1999**, *357*, 1589–1601. [CrossRef]
26. Roters, F.; Eisenlohr, P.; Hantcherli, L.; Tjahjanto, D.; Bieler, T.; Raabe, D. Overview of constitutive laws, kinematics, homogenization and multiscale methods in crystal plasticity finite-element modeling: Theory, experiments, applications. *Acta Mater.* **2010**, *58*, 1152–1211. [CrossRef]
27. Renversade, L.; Quey, R. Intra-grain orientation distributions in deformed aluminium: Synchrotron X-ray diffraction experiment and crystal-plasticity finite-element simulation. *Acta Mater.* **2023**, *262*, 119419. [CrossRef]
28. Zhou, X.; Zan, S.; Zeng, Y.; Guo, R.; Wang, G.; Wang, T.; Zhao, L.; Chen, M. Comprehensive study of plastic deformation mechanism of polycrystalline copper using crystal plasticity finite element. *J. Mater. Res. Technol.* **2024**, *30*, 9221–9236. [CrossRef]
29. Depriester, D.; Goulmy, J.; Barrallier, L. Crystal Plasticity simulations of in situ tensile tests: A two-step inverse method for identification of CP parameters, and assessment of CPFEM capabilities. *Int. J. Plast.* **2023**, *168*, 103695. [CrossRef]
30. Shen, P.; Liu, Y.; Zhang, X. Crystal Plasticity Finite Element Modeling of the Influences of Ultrafine-Grained Austenite on the Mechanical Response of a Medium-Mn Steel. *Crystals* **2024**, *14*, 405. [CrossRef]
31. Frankus, F.; Pachaury, Y.; El-Azab, A.; Devincre, B.; Poulsen, H.F.; Winther, G. Investigating the formation of a geometrically necessary boundary using discrete dislocation dynamics. *J. Mech. Phys. Solids* **2025**, *199*, 106069. [CrossRef]
32. Yuan, T.; He, N.; Ma, S.; Cheng, Y.; Jiang, Y.X.; He, L.; Chen, C.; Chen, Y.; Ye, H. Atomic structure and molecular dynamics simulation of a symmetrical tilt $[0\bar{1}1](511)\Sigma 27$ grain boundary in polysynthetically twinned TiAl crystals. *Mater. Charact.* **2025**, *227*, 115275. [CrossRef]
33. Qin, X.; Liang, Y.; Gu, J. Effects of Stress State, Crack—γ/γ' Phase Interface Relative Locations and Orientations on the Deformation and Crack Propagation Behaviors of the Ni-Based Superalloy—A Molecular Dynamics Study. *Crystals* **2023**, *13*, 1446. [CrossRef]
34. Zhang, Z.; Yang, Y.; Gao, Y.; Wang, G.; Shi, W. Performance Analysis of 7075 Aluminum Alloy Strengthened by Cavitation Water Jet Peening at Different Scanning Speeds. *Crystals* **2022**, *12*, 1451. [CrossRef]
35. Ding, X.; Ma, S.; Zhang, J.; Jiang, Z.; Li, H.; Wang, S.; Wang, C.; Zhong, J. Numerical Simulation and Process Study on Laser Shock Peening of 1Cr18Ni9Ti Material. *Crystals* **2023**, *13*, 1279. [CrossRef]
36. Abdi, M.; Ebrahimi, R.; Bagherpour, E. Improvement of Hydrogenation and Dehydrogenation Kinetics of As-Cast AZ91 Magnesium Alloy via Twin Parallel Channel Angular Extrusion Processing. *Crystals* **2022**, *12*, 1428. [CrossRef]
37. Zhang, H.; Gu, H.; Ma, D. Experimental Study on Laser Lap Welding of Aluminum–Steel with Pre-Fabricated Copper–Nickel Binary Coating. *Crystals* **2025**, *15*, 300. [CrossRef]

38. Chen, Q.; Yu, Y.; Ma, G.; Sun, X.; Lu, L. Dry Sliding Wear Behavior and Mild–Severe Wear Transition of the AA2195-T6 Alloy under Different Loads. *Crystals* **2023**, *13*, 698. [CrossRef]
39. Wang, X.; Du, Z.; Zhang, F.; Zhu, Y.; Liu, Y.; Wang, H. Plastic Damage Assessment in 316 Austenitic Steel Using the Misorientation Parameters from an In Situ EBSD Technique. *Crystals* **2022**, *12*, 1126. [CrossRef]
40. Chang, T.-Y.; Vandenbroeder, G.; Frazer, D.M.; Yushu, D.; Pitts, S.; Chen, T. Nanoindentation Stress Relaxation to Quantify Dislocation Velocity–Stress Exponent. *Crystals* **2024**, *14*, 680. [CrossRef]
41. Liu, Y.; Zhang, Q.; Ge, Q.; Wang, X.; Shen, Y. Improving Texture Prediction by Increasing Mesh Resolution in Submodel: A Crystal Plasticity FE Study and Experiment Verification. *Crystals* **2023**, *13*, 849. [CrossRef]

crystals

MDPI

Article

Crystal Plasticity Finite Element Modeling of the Influences of Ultrafine-Grained Austenite on the Mechanical Response of a Medium-Mn Steel

Pengfei Shen [1], Yang Liu [2] and Xiang Zhang [1,*]

1 Mechanical Engineering, University of Wyoming, Laramie, WY 82071, USA; pshen@uwyo.edu
2 School of Engineering, University of Leicester, Leicester LE1 7RH, UK; yl832@leicester.ac.uk
* Correspondence: xiang.zhang@uwyo.edu

Abstract: Medium manganese (medium-Mn) steel, one of the third-generation advanced high-strength steels (AHSS), delivers impressive mechanical properties such as high yield strength, ultimate tensile strength, and uniform elongation. One notable feature of medium-Mn steels is the presence of ultrafine-grained (UFG) austenite, achieved through phase transformation from the parent martensite phase during intercritical annealing. While, in general, UFG is considered a strengthening mechanism, the impact of UFG austenites in medium-Mn steel has not been fully studied. In this manuscript, we advance our previous work on crystal plasticity simulation based on the Taylor model to consider fully resolved high-fidelity microstructures and systematically study the influence of the UFG austenites. The original microstructure with UFG is reconstructed from a set of serial electron backscatter diffraction (EBSD) scans, where the exact grain morphology, orientation, and phase composition are preserved. This microstructure was further analyzed to identify the UFG austenites and recover them to their parent martensite before the intercritical annealing. These two high-fidelity microstructures are used for a comparative study using dislocation density-based crystal plasticity finite modeling to understand the impact of UFG austenites on both the local and overall mechanical responses.

Keywords: medium-Mn steel; microstructure reconstruction; crystal plasticity finite element modeling; ultrafine-grained austenite

1. Introduction

Advanced high-strength steels (AHSS) deliver impressive mechanical properties such as high yield strengths, ultimate tensile strengths, and uniform elongations, hence playing a crucial role in various engineering applications, especially the automotive industry. AHSS recently started the commercialization of this third-generation product, including medium manganese steels (medium-Mn), designed to combine the economic advantages of the first-generation AHSS with the superior mechanical characteristics of the second-generation [1–3]. Medium-Mn steels are notably characterized by ultrafine-grained (UFG) austenite in their microstructure, produced in the intercritical annealing process at a certain temperature range [4,5]. This temperature range allows the formation of UFG austenite and ferrite within the $\alpha + \gamma$ phase and is also critical for the recrystallization of α'-martensite. The size, quantity, composition, and stacking fault energy of the UFG austenite, therefore, are influenced not just by the partitioning of Mn and C but also by the temperature and duration of intercritical annealing [3,4,6,7].

Among various methods of strengthening, grain refinement is widely acknowledged as an efficient way to enhance strength and toughness simultaneously [8]. For medium-Mn steels, there exists a strong interest in understanding the effects of the UFG phases, which could be used to guide the manufacturing process to improve the properties of medium-Mn steels further. Utilizing approaches like Mileiko's composite model to study the phase fractions of austenite and α'-martensite, Matlock et al. and Gibbs et al. have identified promising enhancements in the mechanical properties of third-generation AHSS [1]. Various experimental studies have identified the contributions to combined strength-ductility from the UFG austenite [9,10], and motivated the manufacturing process design to achieve the desired UFG austenite in the microstructure [11–13]. Crystal Plasticity Finite Element (CPFE) modeling has long been used to establish the microstructure-property relationship for metals and alloys and help guide the analysis and design of better alloys for different applications. A growing trend in AHSS development is the application of CPFE, focusing on studying the deformation and failure mechanisms in multi-phase steels [6,14–16] by incorporating microstructural data, adopting dislocation density-based flow rules and hardening rules for dislocation slip consideration, and accounting for additional deformation mechanisms.

Many medium-Mn steels demonstrate both the twinning induced plasticity (TWIP) effect and the transformation induced plasticity (TRIP) effect, influenced by stacking fault energy [17,18], with strong grain orientation and morphology dependency [19,20]. In our previous work [6], a CPFE modeling framework based on the Taylor model for medium-Mn is developed, which fully incorporates dislocation slip, TWIP, and TRIP but only considers the volume fraction of the individual phases but not the fully resolved microstructures. The impact of UFG is studied by controlling the volume fraction and its evolution in the microstructure in the context of the Taylor model, where each grain/phase is subjected to the same macroscopic loading and the volume fraction weighted average response is used as the macroscopic response. The TRIP effect in medium-Mn involves the transformation of retained austenite into martensite during plastic deformation [21]. This transformation is associated with a significant transformation strain, which can lead to plastic strains of up to 35% without compromising strength. Mechanical twinning, a common phenomenon in many metallic materials, including medium-Mn, facilitates material deformation through the simultaneous cooperative shear of numerous planes, enabling large shear strains. Both TRIP and TWIP significantly increase the work hardening rate and strengthen the material [4,21–23]. Previous studies on medium-Mn indicate that mechanical twinning was observed at an early age of 8% in engineering strains, and it is assumed that austenite did not transform before this strain [6]. In addition, seeing the increasing trend in using medium-Mn in cases where the deformation is expected to be lower than the threshold for TRIP and TWIP and the promising potential achieved by dislocation slip alone [24], we consider dislocation slip as the sole deformation mechanism in our current study but adopt fully resolved microstructures that explicitly represent different phases and individual grains, including the UFG, to study the effect of the UFG on the mechanical response of medium-Mn systematically.

The remainder of this manuscript is organized as follows: Section 2 introduces the flow and hardening rules used in the CPFE model, followed by the details on microstructure reconstruction, meshing, and phase identification of medium-Mn steel in Section 3. A comparative study between two microstructures with or without UFG austenites to investigate the effects of the UFG austenite on the mechanical response is discussed in Section 4. Section 5 summarizes and concludes the manuscript.

2. CPFE Modeling of Medium-Mn: Dislocation Density-Based Flow and Hardening Rules

CPFE relies on fully resolved polycrystalline microstructures and relevant deformation mechanisms, enabling a detailed study of the microscale response [25]. While medium-Mn alloys feature TWIP and TRIP, as studied in our previous study, where a Taylor-based homogenization scheme is used [6], we consider fully resolved microstructure but only deformation slip as the deformation mechanism, relevant to the increasing interest in using medium-Mn under small deformation as described above. To describe the dynamics of dislocation slip under varying stress and thermal conditions, we consider the flow rule of [26,27]

$$\dot{\gamma}^{\alpha} = \frac{\rho_{\mathrm{m}} \mathrm{v}_{\mathrm{id}} \mathrm{b}^{\alpha 2}}{2} \exp\left(\frac{-\Delta \mathrm{F}}{\mathrm{k_B} \vartheta}\right) \exp\left(\frac{(\tau^{\alpha} - \mathrm{s}^{\alpha}) \Delta \mathrm{V}}{\mathrm{k_B} \vartheta}\right) \mathrm{sgn}(\tau^{\alpha}), \quad (1)$$

where $\dot{\gamma}^{\alpha}$ represents the shear strain rate associated with dislocation slip in the αth slip system. The shearing becomes active when the resolved shear stress (RSS) τ^{α} surpasses its critical value, the critical resolved shear stress (CRSS) S^{α}. Otherwise, the slip rate is zero. ρ_m represents the average mobile dislocation density, and ΔF denotes the Helmholtz free energy, which acts as the effective activation energy barrier for dislocation glide. The magnitude of the Burgers vector is represented by b, and v_{id} signifies the dislocation vibration frequency. ϑ denotes the temperature. Additionally, k_B is the Boltzmann constant, and ΔV represents the thermal activation volume.

The slip resistance S^{α} has its contributions from four distinct sources: the initial temperature associated S_0^{α} [28,29], Hall-Petch associated S_{HP}^{α} [30], forest dislocation density related S_{for}^{α} [31], and substructure related S_{deb}^{α} [32], described by [6,33]

$$s^{\alpha} = s_0^{\alpha} + s_{\mathrm{HP}}^{\alpha} + s_{\mathrm{for}}^{\alpha} + s_{\mathrm{deb}}^{\alpha}, \tag{2}$$

where,

$$s_{\mathrm{HP}}^{\alpha} = \mu \mathrm{HP}^{\alpha} \sqrt{\frac{\mathrm{b}^{\alpha}}{\mathrm{d}_{\mathrm{g}}^{\alpha}}}, \tag{3}$$

$$s_{\mathrm{for}}^{\alpha} = b^{\alpha} \chi \mu \sqrt{\rho_{\mathrm{for}}^{\alpha}}, \tag{4}$$

$$s_{\mathrm{deb}}^{\alpha} = \mathrm{k}_{\mathrm{deb}} \mu \mathrm{b}^{\alpha} \sqrt{\rho_{\mathrm{deb}}} \ln\left(\frac{1}{\mathrm{b}^{\alpha} \sqrt{\rho_{\mathrm{deb}}}}\right). \tag{5}$$

In the above, χ acts as the dislocation interaction constant, and μ is the elastic shear modulus. The forest dislocation density within an individual grain is represented by ρ_{for}^{α}. This study sets a material-independent constant, k_{deb}, to 0.1 for the three phases, highlighting a standardized approach to handling different crystallographic structures [34]. Furthermore, ρ_{deb} signifies the total debris dislocation density for a single grain, a crucial factor in determining the shear resistance. The Hall-Petch associated coefficient, HP^{α}, is another key element in the equation, and it interplays with d_g^{α}, the grain size of each corresponding phase. This approach allows for a nuanced and detailed analysis of shear resistance, considering a variety of factors that influence it in different phases and under various conditions. The hardening is captured through the evolution of ρ_{for}^{α} and ρ_{deb}^{α} by a set of evolution equations following Refs. [6,32,35]. The evolution of forest dislocation is described by

$$\frac{\partial \rho_{\mathrm{for}}^{\alpha}}{\partial \gamma^{\alpha}} = \frac{\partial \rho_{\mathrm{gen}}^{\alpha}}{\partial \gamma^{\alpha}} - \frac{\partial \rho_{\mathrm{rec}}^{\alpha}}{\partial \gamma^{\alpha}} = k_1 \sqrt{\rho_{\mathrm{for}}^{\alpha}} - k_2(\dot{\gamma}, \vartheta) \rho_{\mathrm{for}}^{\alpha}, \tag{6}$$

$$k_2(\dot{\gamma}, \vartheta) = k_1 \frac{b^{\alpha} \chi}{g^{\alpha}} \left[1 - \frac{k\vartheta}{\hat{D} b^3} \ln\left(\frac{\dot{\gamma}}{\dot{\gamma}_0}\right)\right], \tag{7}$$

where k_1 represents an adjustable coefficient related to the generation of statistically stored dislocations due to forest trapping of mobile dislocations; k_2 is the rate-sensitive coefficient for dynamic recovery. g^{α} denotes the normalized stress-independent activation energy, $\dot{\gamma}_0$ is the reference strain rate, and $\hat{D}$ is the drag stress [32,36].

The debris dislocation density evolution captures the hardening due to substructure and is defined as the accumulative effects from multiple slip systems for one grain [32]

$$d\rho_{\mathrm{deb}}^{(\alpha)} = \sum_s \frac{\partial \rho_{\mathrm{deb}}^{s(\alpha)}}{\partial \gamma^{s(\alpha)}} d\gamma^{s(\alpha)} = \sum_s q b^{s(\alpha)} \sqrt{\rho_{\mathrm{deb}}^{(\alpha)}} k_2^{s(\alpha)} \rho_{\mathrm{for}}^{s(\alpha)}, \tag{8}$$

in which q^{α} is a recovery rate coefficient, and the initial dislocation debris densities in all slip systems are defined to be the same value of $\rho_{\mathrm{deb},0}^{s(\alpha)} = 1.0 \times 10^{10}\ \mathrm{m}^{-2}$ [37,38].

Equations (1)–(8) complete the constitutive law description of medium Mn, considering the plastic deformation accumulation via dislocation slip and hardening behavior through the evolution of different types of dislocation densities. This set of constitutive equations, together with a finite strain kinematics description using the multiplicative decomposition [39,40], allows the implementation as a user material subroutine (UMAT) in the finite element software Abaqus 2019.

3. Microstructure Reconstruction and Phase Identification

The medium-Mn alloy studied in this work is manufactured from a Fe-based alloy with mass fractions of 12 wt% Mn, 3 wt% Al, and 0.05 wt% C after going through hot rolling at 900 °C, cold rolling, and intercritical annealing at 585 °C for 8 h [3,6]. Serial Electronic Backscatter Diffraction (EBSD) scanning is conducted to characterize the microstructure at a step size of 50 nm, following the detailed procedure reported in [3,6]. All the EBSD data, together with the experimental tensile stress-strain response of the same materials from these prior studies, are adopted in the current work.

The open-source software DREAM.3D [41] (version 6.5) is adopted to process the EBSD data and reconstruct the microstructure. DREAM.3D can either process the EBSD data to extract the statistical information (e.g., grain size and orientation distribution) and synthesize statistically equivalent microstructures of different sizes [42,43], or reconstruct exact microstructures using single (2D) or multiple EBSD scans [44]. We adopted the second method and used the EBSD data from the previous study in Refs. [3]. In this process, as shown in Figure 1, the individual EBSD scan is first cleaned and analyzed to identify different phases based on the lattice structure and individual grains based on different orientations using a misorientation angle threshold of 5°. The neighboring scans are then aligned and stacked together to reconstruct voxel-based 3D microstructures, which can be further meshed with hexagonal elements by changing each voxel into one element.

(a) (b) (c)

Figure 1. The process of microstructure reconstruction involves: (**a**) processing individual layers; (**b**) stacking all layers and synthesizing an exact microstructure; and (**c**) generating a finite element model.

In medium-Mn steels, austenite is characterized by a face-centered cubic (FCC) crystal structure. At the same time, ferrite is of body-centered cubic (BCC) structure, and α'-martensite typically has a body-centered tetragonal (BCT) structure referred to as BCT-martensite. EBSD differentiates phases based on their crystal lattice structures, making it straightforward to identify austenite. However, identifying low-carbon α'-martensite in the presence of ferrite is challenging with EBSD. This difficulty arises because the martensite's low tetragonality makes its lattice structure very similar to that of BCC ferrite, which is below EBSD's indexing capability. As a result, both ferrite (BCC) and α'-martensite (BCT) are often detected as the same BCC phase in EBSD and analyzed together [3,45,46].

To precisely distinguish ferrite and martensite phases in medium-Mn steels, both the scanning transmission electron microscopy via energy-dispersive spectroscopy (STEM-EDS) and X-ray diffraction (XRD) analysis have been used in the literature [47,48]. However, it remains a great challenge to combine these techniques with serial EBSD so that a fully resolved polycrystalline microstructure with precise phase information can be reconstructed. Alternatively, leveraging the reported grain size difference between ferrite and austenite, one could separate the two phases in an approximated fashion following the procedures in Figure 2.

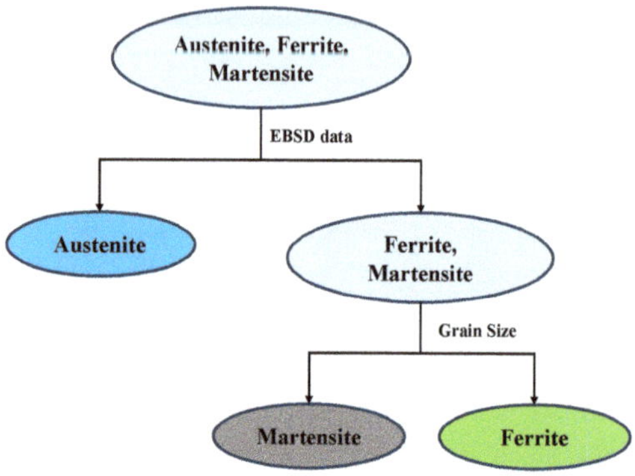

Figure 2. Phase identification procedure: austenite is distinguished directly by EBSD data, while martensite and ferrite are then recognized according to grain size distribution.

In a recent study on the same material as in this work, by examining the grain size distribution (in terms of grain area) of the combined ferrite and martensite phases in a single EBSD scan, a bimodal distribution was observed [3]. This distribution features a sharp peak in grain area at 79.1 μm^2 and a smaller peak at 3.58 μm^2, respectively, representing the martensite and ferrite phases. The corresponding grain size distribution, converted from the grain area distribution, under the equiaxed grain shape assumption, is plotted in Figure 3a. It should be noted that the actual grains do not follow an equiaxed grain shape, but this is a necessary approximation to separate the two phases by size and is sufficient for the qualitative study in the current work. The grain size distribution shows a high area fraction between the two peaks, making it difficult to accurately separate the two phases by size. We further assume 2.90 μm as a shart threshold to separate the two different phases. This is done by computing the sizes of all the grains in the microstructure except the austenites and setting all grains with grain sizes smaller than an equiaxed grain with size 2.90 μm as ferrite, while setting those larger than this as martensite. Using this phase size threshold, the phase distribution of the fully resolved microstructure is shown in Figure 3b.

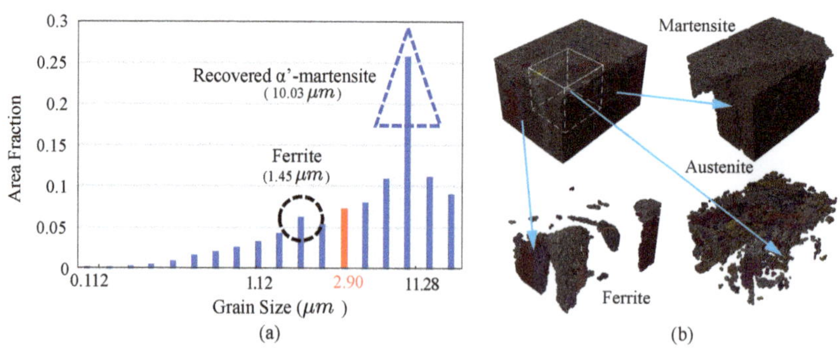

Figure 3. (**a**) Grain size distribution of martensite and ferrite computed from the grain area distribution in Ref. [3]; (**b**) separated austenite, ferrite, and martensite phases in the microstructure. The white block demonstrates the subdomain that will be used later.

3.1. Mesh Size Convergence Study and Model Calibration

When reconstructing the microstructure in DREAM.3D, as described in Section 3, the user can specify the reconstruction resolution to be used. The smallest resolution one can use is the EBSD resolution of 50 nm (i.e., EBSD step size), which will identify 1473 grains in the reconstructed microstructure (size $9.0 \times 6.45 \times 5.45$ μm), with a total of 2,530,980 elements and 2,482,194 nodes by meshing each voxel into one 8-node hexahedral element (C3D8) as shown in Figure 3b. However, the large number of elements and nodes in this model make it computationally expensive; hence, a mesh size sensitivity study is carried out to identify the proper mesh size to better balance the computational cost and accuracy.

A series of gradually coarsened meshes is reconstructed using the reconstruction resolutions between 0.05 and 1.0 μm and meshing each pixel into one element (i.e., mesh size is equal to the reconstruction resolution). The corresponding mesh information is reported in Table 1. The reconstructed microstructures are illustrated in Figure 4a, while the stress-strain responses (using the parameters from Ref. [6]) are plotted in Figure 4b. For the martensite, BCC slip systems are used to approximate the BCT slip systems due to the low carbon content, which is expected to have a negligible change in the lattice parameters [49].

Table 1. Model size comparison when using different reconstruction resolutions.

Recon. resolution (μm) Mesh size (μm)	1 1.0	0.5 0.5	0.25 0.25	0.1 0.1	0.08 0.08	0.05 0.05
No. of grains	26	184	491	1170	1248	1473
No. of C3D8 elements	270	2160	18,900	311,040	609,280	2,530,980
No. of nodes	601	4067	29,131	381,803	696,857	2,482,194

It is clearly shown that when a large reconstruction resolution is used, small grains are lost in the reconstruction process, leading to a significant difference compared to the finest model. As the reconstruction resolution decreases, the small grains are better preserved. When a resolution of 0.08 μm is used, the stress-strain response is almost identical to that from the 0.05 μm reconstruction resolution but with a much smaller number of elements and nodes. Hence, 0.08 μm is considered an optimal mesh size to use for balanced computational cost and accuracy.

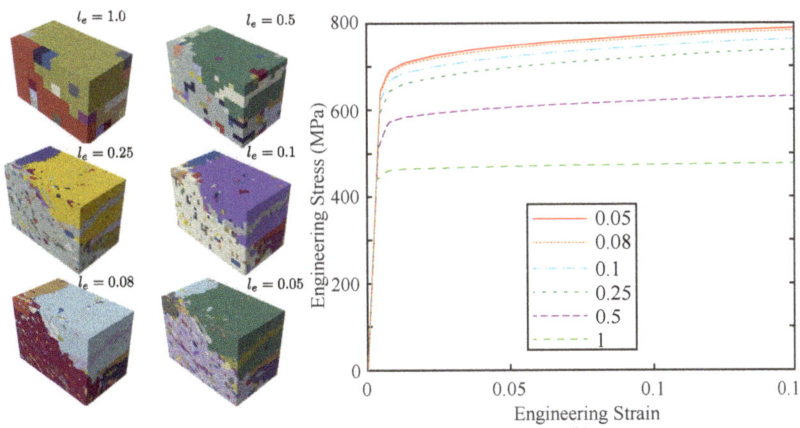

Figure 4. Mesh size convergence study: (**a**) reconstructed microstructures using a resolution of 1.0, 0.5, 0.25, 0.1, 0.08, and 0.05 μm; (**b**) stress-strain responses from different mesh sizes.

While the material parameters start from our previous work [6], additional fine-tuning is conducted to match the experimental stress-strain curve from previous work [3,6]. This fine-tuning is needed since the current study uses a fully resolved microstructure, while the previous work considered the phase volume fraction only when using the Taylor model. Specifically, three different strain rates are considered, including a low strain rate (LS, 2×10^{-4}/s), a medium strain rate (MS, 2×10^{-2}), and a high strain rate (HS, 2×10^{1}/s), using the mesh size of 0.08 μm. The calibration process yields a good match between the experiments and simulations across the three different strain rates as shown in Figure 5. The slightly adjusted material parameters are provided in the Appendix A for simplicity.

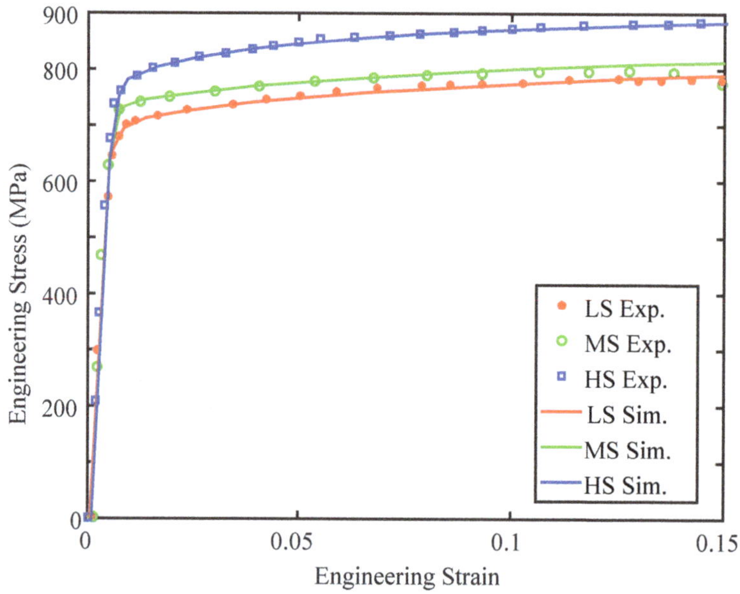

Figure 5. Stress-strain curve of the simulation compared to the experimental data.

3.2. Establish a Microstructure without Austenite

To specifically assess the influence of austenite, a comparative analysis between two microstructures with and without austenite is essential. This necessitates the creation of an alternative microstructure that excludes UFG austenites. Since austenites are typically derived from martensite during the heat treatment process when the alloy is produced, we develop a strategy to identify the parent martensite of each UFG austenite and convert those austenites to their parent martensite before the intercritical annealing. By converting these identified austenite grains back into martensite, we can develop a comparative model that effectively excludes austenite, allowing for a more focused study on the influence of the austenite on the mechanical response of the medium-Mn steel.

Given that austenite originates from a parent martensite, each austenite grain is typically encompassed by martensite grains. Therefore, the approach is to identify the neighboring martensite (parent martensite) for each UFG austenite and change each austenite to its parent martensite. The neighboring information is obtained by searching the shared grain boundary nodes between two grains. When an austenite grain is neighboring multiple martensite grains, the martensite that shares the maximum number of grain boundary nodes is deemed the parent martensite of that specific austenite. If an austenite grain has the same number of neighboring nodes with more than one martensite grain, it will be randomly changed to one of the martensite grains. Figure 6 illustrates the procedure, where the initial four steps in (a)–(d) outline the microstructure reconstruction process and the identification of three distinct phases. Steps (e) and (f) illustrate the process of determining the parent martensite of the austenite and recovering the austenite to its parent martensite.

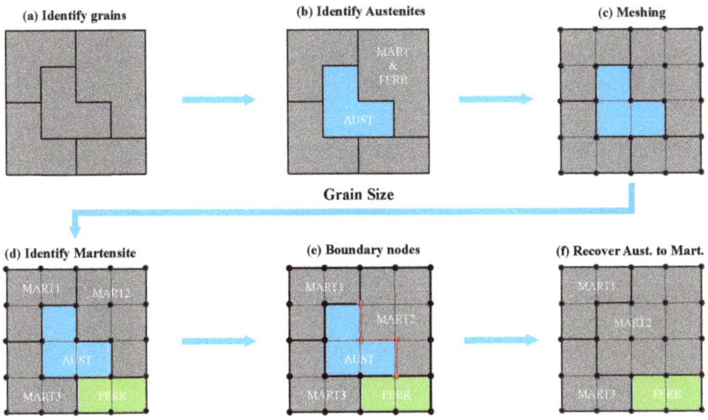

Figure 6. Process of recognizing austenite, ferrite, and martensite phases and recovering austenites to their parent martensite phase: (**a**) grain, (**b**) austenite identification, and (**c**) meshing in DREAM3D; (**d**) separating ferrite from martensite by grain size; (**e**) searching for the shared nodes between each austenite and its surrounding martensite; (**f**) transforming the austenite to the parent martensite, which has the largest number of neighboring nodes with the austenite. Thick black lines indicate grain boundary, thin black lines indicate element boundary, and dots indicate nodes; red dots are the shared nodes between austenite and its parent martensite.

Utilizing this methodology, it becomes possible to create two distinct microstructural models: one that includes austenite and another that represents the microstructure before the intercritical annealing and free to austenites, enabling a comprehensive comparison and analysis. However, due to the large number of elements in the generated finite element model, which would result in a time-consuming simulation process, we, therefore, select a subdomain extracted from the entire model for more efficient analysis, as shown in Figure 7, and further use an element size of 1.0 µm since it provides close enough results compared to the 0.08 µm mesh size, but with a significantly smaller number of elements and nodes (e.g., see Table 1). This particular subdomain is chosen because it contains a cluster of UFG austenites (9% volume fraction of austenite), and its location in the whole microstructure is shown in Figure 3a. The two models are shown in Figure 8, and we will call these two models the model with austenite and the model without austenite.

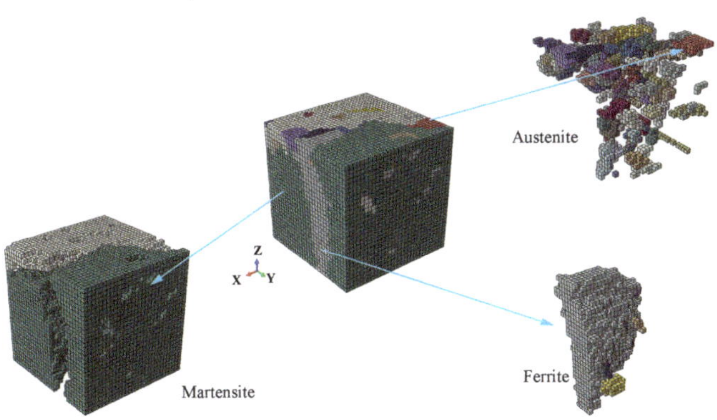

Figure 7. Subdomain chosen from the entire microstructure.

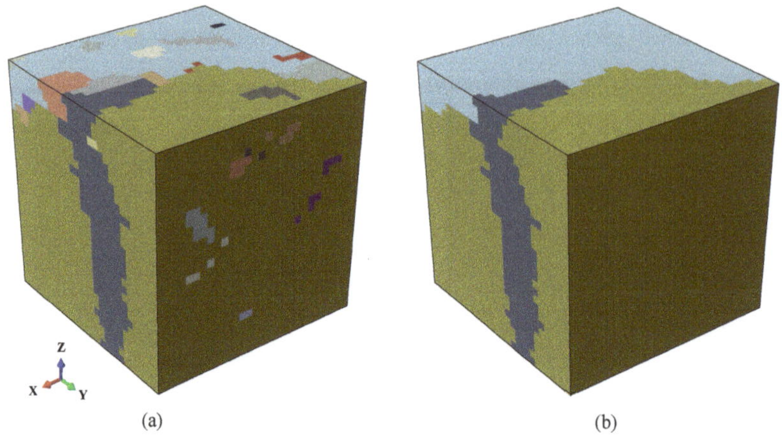

Figure 8. Finite element models for the chosen subdomain: (**a**) the model with austenite, ferrite, and martensite; (**b**) the comparative model with all austenites transformed back to corresponding parent martensite.

4. CPFE Simulation Results and Discussion

We now focus on understanding the impact of the UFG austenite by running simulations on the microstructures with and without austenite in Figure 8. We consider uniaxial tensile loading in the X direction up to 10% strain. The boundary conditions are shown in Figure 9a, where all nodes on the $-X$, $-Y$, and $-Z$ faces are fixed in their normal directions, while nodes on the $+X$ faces are displaced in the $+X$ direction. Figure 9b shows the engineering stress-strain curves simulated from the original and modified microstructures. The stress-strain curve is different from that in Figure 5, mainly because the two big austenites span over the entire microstructure in the X direction, carrying a large amount of the applied strain. The model incorporating austenite displays a curve that is slightly elevated compared to the model devoid of austenite. This suggests that the presence of austenite contributes to an increase in both the yield stress and strength. A prior study using the Taylor model [6] suggested that by increasing the UFG austenite's volume fraction from 35% to 70%, the ultimate tensile strength changes from 700 to 900 MPa, a 22% increase. In the current study, the volume fraction of UFG austenite for this example is 9%, and the stress at 6% strain has a 3% increase.

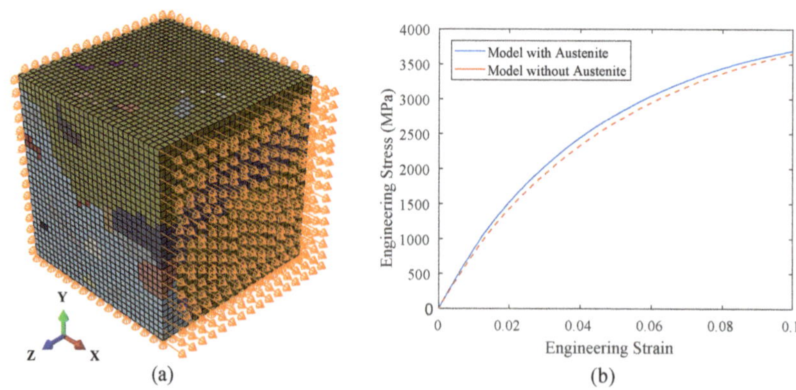

Figure 9. (**a**) Boundary conditions of the finite element model; (**b**) Stress-strain curves of the model with and without austenites.

Figure 10 shows the contour of stress component σ_{11} in the models with or without austenite at 50% of the applied load. It should be noted that while the stress distributions on the $+X$ face might be influenced by the boundary conditions, ferrite generally exhibits a lower stress level. When comparing models with and without austenite, an apparent high stress level is observed in the austenite region, suggesting austenite's contribution to the increased strength.

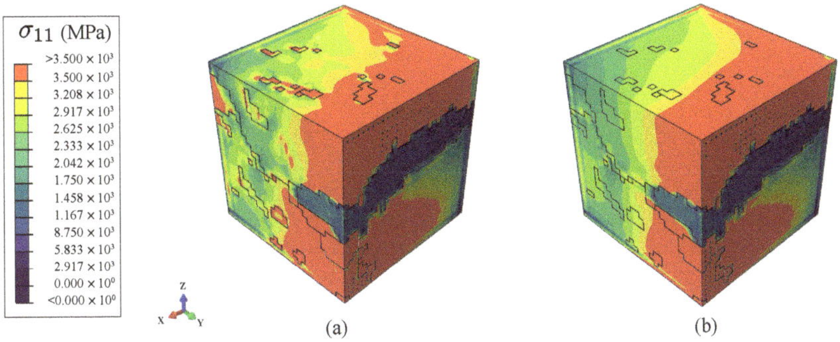

Figure 10. Distribution of stress σ_{11} at 50% of the applied loading for (**a**) the model with austenite and (**b**) the model without austenite.

The distributions of true strain component ϵ_{11}^{T} and equivalent plastic strain ε^{ep} are shown in Figures 11 and 12, respectively. The equivalent plastic strain is defined as [50,51]

$$\varepsilon^{eP} = \sqrt{\frac{2}{3}\varepsilon_{ij}^{p}\varepsilon_{ij}^{p}} \tag{9}$$

where ϵ_{ij}^{p} is the plastic strain using index notation, obtained by time integration of the plastic strain rate [52,53]. Similarly, while the boundary conditions may impact the strain distribution on the $+X$ face, higher plastic strain levels are observed in the austenite areas, suggesting its contribution to the ductility. In addition, high true strain is also observed in the martensite neighboring some of the austenite grains, indicating a strong interaction between the austenite and its neighboring martensite. Overall, ferrite exhibits significantly higher levels of both true and plastic strains compared to austenite and martensite, with martensite showing the lowest level of plastic strain. This observation aligns with experimental findings [6].

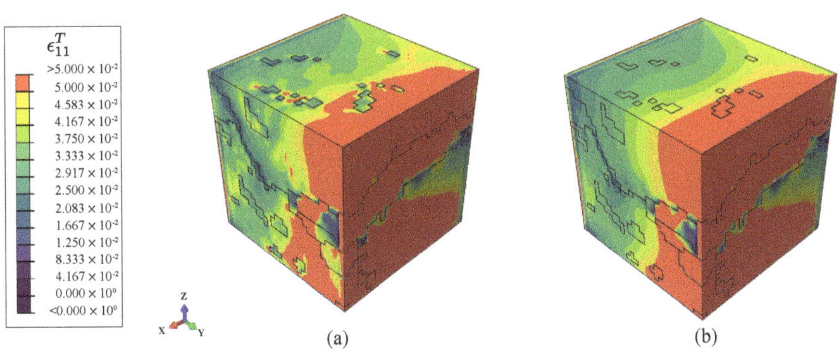

Figure 11. Distribution of true strain ϵ_{11}^{T} at 50% of the applied loading for (**a**) the model with austenites and (**b**) the model without austenites.

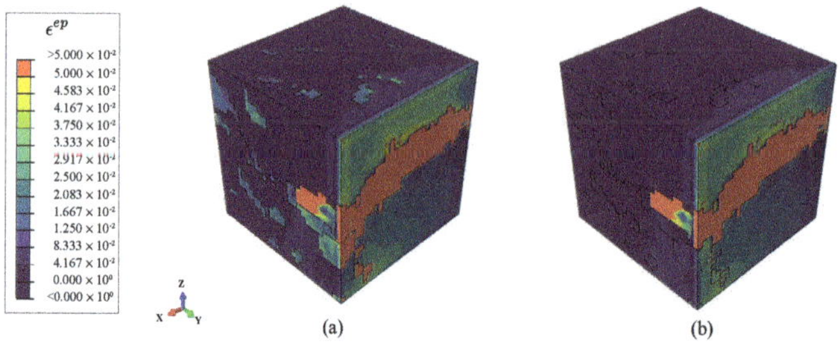

Figure 12. Distribution of equivalent plastic strain ϵ^{ep} at 50% of the applied loading for (**a**) the model with austenites and (**b**) the model without austenites.

Figures 13b–d illustrate the values of stress component σ_{11}, true strain component ϵ^{T}_{11}, and equivalent plastic strain ϵ^{ep}, respectively, along a specific path in the microstructure. This path was created at the same position in both models, as shown in Figure 13a, starting from a martensite and crossing an austenite and a ferrite. The figures demonstrate that the responses for models with and without austenites are largely similar, with the notable exception of the austenite regions. In these regions, the curve representing the model with austenite exhibits much higher stress (about 25% higher) and plastic strain (about 50% higher) values compared to the model without austenite but only marginally lower true strain. These observations from the inside of the microstructure are consistent with the observations from the surface of the microstructure shown in the contour plots of Figures 10–12. This again confirms that austenite carries much greater plastic deformation and higher stresses, which could significantly contribute to ductility and high strength, making it a valuable component in the material's overall performance.

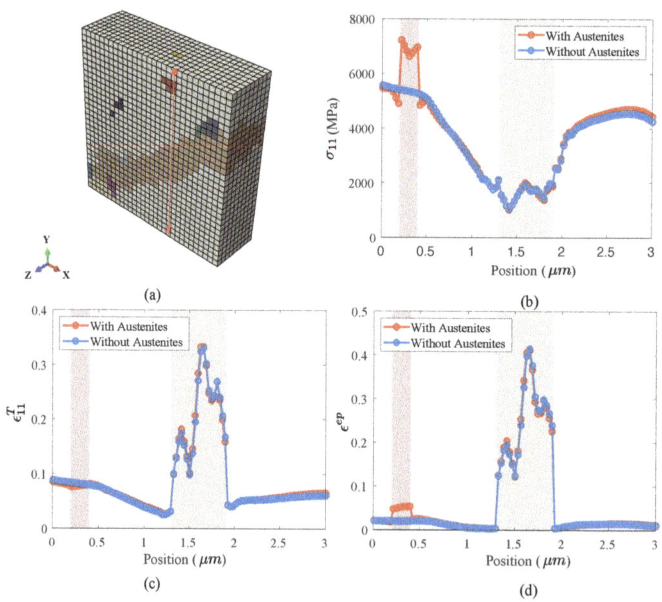

Figure 13. Local responses over the path across martensite, austenite, and ferrite at the end of the loading: (**a**) the path passing the martensite (yellow area), austenite (red area), and ferrite (orange area). Response along the path includes (**b**) stress component σ_{11}, (**c**) true strain component ϵ^{T}_{11}, and (**d**) equivalent plastic strain ϵ^{ep}. The narrower shaded region in (**b**–**d**) indicates the austenite, while the wider shaded region indicates the location of the ferrite.

To understand the statistical aspects of the response, Figure 14 shows the histograms of stress component σ_{11}, true strain component ϵ_{11}^{T}, and equivalent plastic strain ϵ^{ep}. The left column in Figure 14 shows the results of the model with austenite, while the right column shows the model without austenite. The X axis refers to the value of local responses, while the Y axis represents the number of integration points that have the corresponding response value. The histogram for σ_{11} reveals that most ferrite grains exhibit lower stress values than those of the other two phases, consistent with the contour and path plots. In σ_{11} results without austenite, where all austenite grains were reverted to their respective martensites, the distributions of the original martensite and recovered martensite (from austenite) follow a similar trend as expected, but the stress in austenite is higher than that in martensite. A comparison of ϵ_{11}^{T} and the equivalent plastic strain ϵ^{ep} in the presence of austenites shows that both martensite and austenite experience similar levels of true strain, while austenite displays a higher plastic strain level. These observations confirm that the prior observation of higher stress and plastic strain in austenite is true for the entire microstructure, and the contribution to the ductility and high strength of medium-Mn steel is true for all austenite grains, not just the particular grains we analyzed previously. Furthermore, the distribution of ϵ_{11}^{T} and equivalent plastic strain ϵ^{ep} in ferrite spans a much broader range and especially very high plastic strain levels. These high plastic strains are much higher than the applied strain, indicating a very strong interaction between ferrite and surrounding grains and its contribution to the ductility of medium-Mn steel.

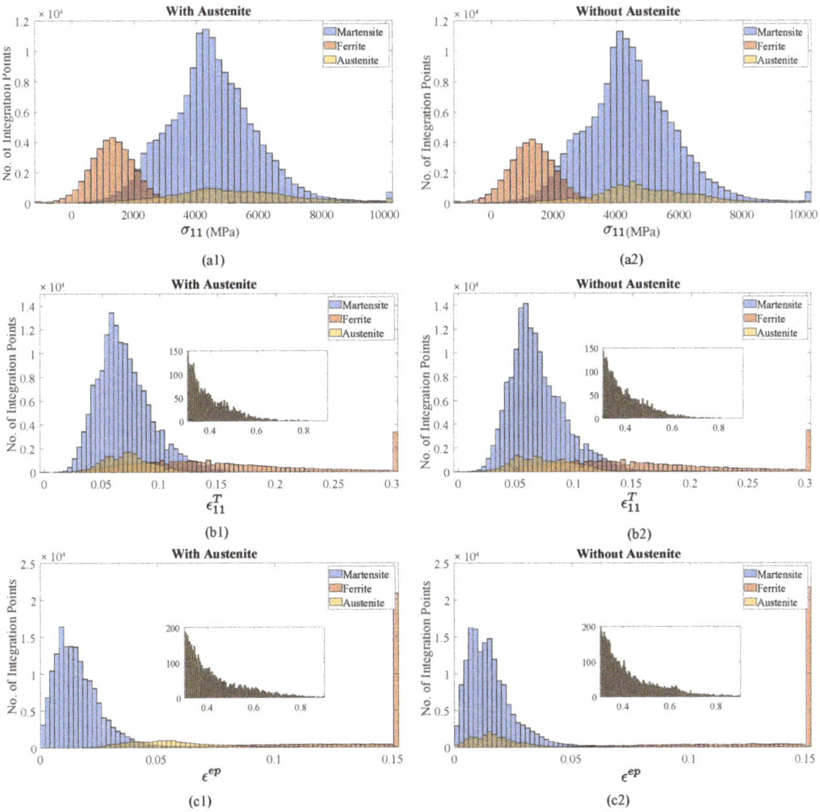

Figure 14. Histogram of stress component σ_{11}, true strain ϵ_{11}^{T}, and equivalent plastic strain ϵ^{ep} at the end of applied loading: (**a1**,**b1**,**c1**) are results of the model without austenite; (**a2**,**b2**,**c2**) are results of the model without austenite.

Figure 15 demonstrates the relationship between the Schmid factor and σ_{11}. Similar to Figure 14, the left column of the figure represents the model that includes austenite, while the right column depicts the model without austenite. It should be noted that the range of the Schmid factor before loading is mainly determined by the number of grains in each phase, and the spreading as loading continues reflects the re-orientation (i.e., texture evolution) during the loading process. Overall, texture evolution is observed for all three phases, but ferrite has the most texture evolution at a relatively low stress level. This further confirms ferrite's strong interactions with neighboring grains and its contribution to medium-Mn's ductility. It is also observed that the presence of austenite does not seem to change the response of ferrite and martensite much.

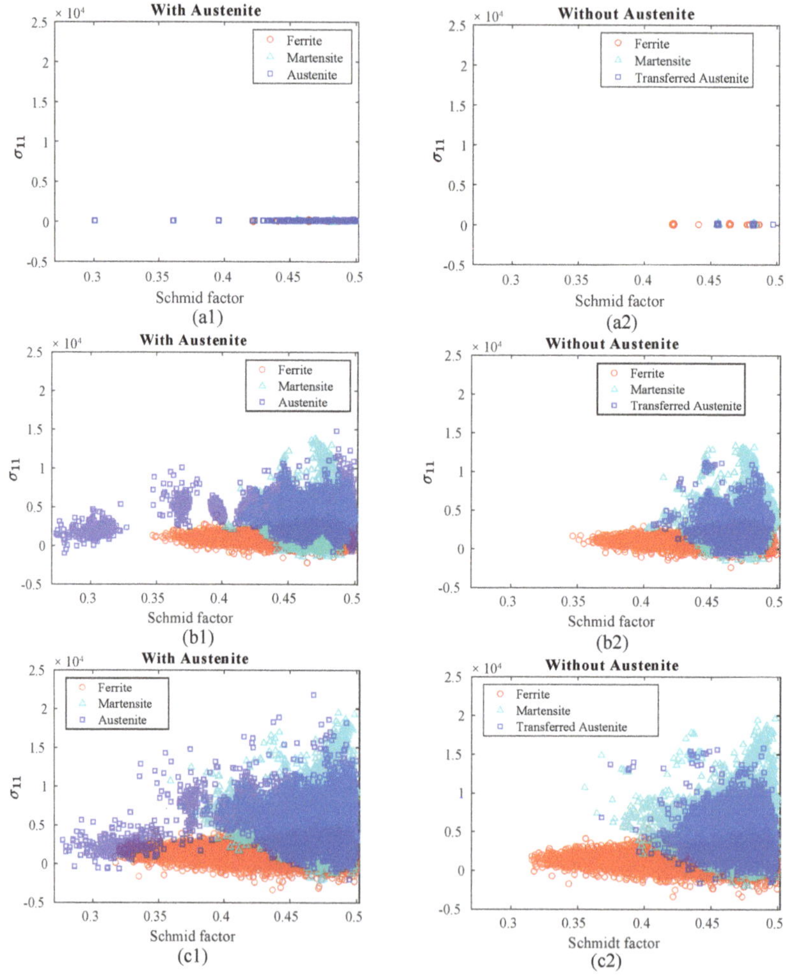

Figure 15. Correlation between Schmid factor and stress component at (**a1**,**a2**) 0% of loading, (**b1**,**b2**) 50% of loading, and (**c1**,**c2**) 100% of loading.

5. Conclusions

We conducted a comparative CPFE study with the microstructures with and without austenite to investigate the influences of UFG austenite on the mechanical response of a medium-Mn steel. The microstructure of austenite is constructed by stacking serial

EBSD scans together in DREAM.3D. The microstructure without austenite is obtained by transforming all austenites to the corresponding parent martensites before the intercritical annealing. We systematically studied the response in terms of the histogram and distribution of σ_{11}, true strain ϵ_{11}^{T} and equivalent plastic strain ϵ^{ep} at the surface of the microstructure, along a patch inside the microstructure, and their statistical distribution over the entire microstructure. The comparison indicates that austenite carries much higher plastic strain and a higher stress level than the other two phases, contributing to the ductility and high strength of medium-Mn steel.

Author Contributions: Conceptualization, Y.L. and X.Z.; methodology, P.S., Y.L. and X.Z.; software, P.S. and Y.L.; validation, P.S. and Y.L.; formal analysis, P.S.; investigation, P.S., Y.L. and X.Z.; resources, X.Z.; data curation, P.S. and Y.L.; writing—original draft preparation, P.S.; writing—review and editing, P.S., Y.L. and X.Z.; visualization, P.S.; supervision, Y.L. and X.Z.; project administration, X.Z.; funding acquisition, X.Z. All authors have read and agreed to the published version of the manuscript.

Funding: This research received no external funding and has been supported by the faculty startup funding at the University of Wyoming.

Data Availability Statement: Data in this work can be available upon reasonable request due to privacy.

Acknowledgments: The authors greatly acknowledge the experimental data and helpful discussions provided by Jake Benzing from the National Institute of Standards and Technology.

Conflicts of Interest: The authors declare no conflict of interest.

Appendix A

The material properties appear in Section 2 builds on our prior study using Taylor model [3], and was further calibrated using the CPFE model in conjunction with the fully resolved microstructure as described in Section 3. The material parameters are provided in Table A1.

Table A1. Material parameters in the CPFE model.

Parameters		Phase		
Symbol	Unit	Martensite	Ferrite	Austenite
C_{11}	GPa	169	175	174
C_{12}	GPa	82	82	85
C_{44}	GPa	96	97	99
ρ_m	m^{-2}	10^{12}	10^{12}	10^{12}
ν_{id}	Hz	10^{13}	10^{13}	10^{13}
b^α	µm	2.50×10^{-4}	2.50×10^{-4}	2.56×10^{-4}
ΔF^α	J	4.45×10^{-20}	4.45×10^{-20}	6.39×10^{-21}
ΔV^α	m^3	7.69×10^{-29}	7.69×10^{-29}	6.25×10^{-28}
S_0^α	MPa	1036.4	272.7	163.4
μ^α	GPa	320	320	207
HP^α	1	130	130	134
k^1	m^{-1}	4.29×10^{7}	4.29×10^{7}	3.51×10^{8}
$\rho_{for,0}$	m^{-2}	10^{10}	10^{10}	10^{14}

References

1. Matlock, D.K.; Speer, J.G. Processing Opportunities for New Advanced High-Strength Sheet Steels. *Mater. Manuf. Process.* **2010**, *25*, 7–13. [CrossRef]
2. Hu, B.; Luo, H.; Yang, F.; Dong, H. Recent progress in medium-Mn steels made with new designing strategies, a review. *J. Mater. Sci. Technol.* **2017**, *33*, 1457–1464. [CrossRef]
3. Benzing, J.T.; Kwiatkowski da Silva, A.; Morsdorf, L.; Bentley, J.; Ponge, D.; Dutta, A.; Han, J.; McBride, J.R.; Van Leer, B.; Gault, B.; et al. Multi-scale characterization of austenite reversion and martensite recovery in a cold-rolled medium-Mn steel. *Acta Mater.* **2019**, *166*, 512–530. [CrossRef]

4. Lee, S.; De Cooman, B.C. Annealing Temperature Dependence of the Tensile Behavior of 10 pct Mn Multi-phase TWIP-TRIP Steel. *Metall. Mater. Trans. A* **2014**, *45*, 6039–6052. [CrossRef]
5. Benzing, J.T.; Bentley, J.; McBride, J.R.; Ponge, D.; Han, J.; Raabe, D.; Wittig, J.E. Characterization of Partitioning in a Medium-Mn Third-Generation AHSS. *Microsc. Microanal.* **2017**, *23*, 402–403. [CrossRef]
6. Benzing, J.T.; Liu, Y.; Zhang, X.; Luecke, W.E.; Ponge, D.; Dutta, A.; Oskay, C.; Raabe, D.; Wittig, J.E. Experimental and numerical study of mechanical properties of multi-phase medium-Mn TWIP-TRIP steel: Influences of strain rate and phase constituents. *Acta Mater.* **2019**, *177*, 250–265. [CrossRef] [PubMed]
7. Han, J.; da Silva, A.K.; Ponge, D.; Raabe, D.; Lee, S.M.; Lee, Y.K.; Lee, S.I.; Hwang, B. The effects of prior austenite grain boundaries and microstructural morphology on the impact toughness of intercritically annealed medium Mn steel. *Acta Mater.* **2017**, *122*, 199–206. [CrossRef]
8. Wang, T.; Hu, J.; Misra, R.D.K. Microstructure evolution and strain behavior of a medium Mn TRIP/TWIP steel for excellent combination of strength and ductility. *Mater. Sci. Eng. A* **2019**, *753*, 99–108. [CrossRef]
9. Hu, J.; Zhang, J.M.; Sun, G.S.; Du, L.X.; Liu, Y.; Dong, Y.; Misra, R.D.K. High strength and ductility combination in nano-/ultrafine-grained medium-Mn steel by tuning the stability of reverted austenite involving intercritical annealing. *J. Mater. Sci.* **2019**, *54*, 6565–6578. [CrossRef]
10. Hu, B.; Rong, X.; Tian, C.; Yu, Y.; Guo, H.; Misra, R.D.K.; Shang, C. Nanoscale precipitation and ultrafine retained austenite induced high strength-ductility combination in a newly designed low carbon Cu-bearing medium-Mn steel. *Mater. Sci. Eng. A* **2021**, *822*, 141685. [CrossRef]
11. Sadeghpour, S.; Somani, M.C.; Kömi, J.; Karjalainen, L.P. A new combinatorial processing route to achieve an ultrafine-grained, multiphase microstructure in a medium Mn steel. *J. Mater. Res. Technol.* **2021**, *15*, 3426–3446. [CrossRef]
12. Jia, Q.; Chen, L.; Xing, Z.; Wang, H.; Jin, M.; Chen, X.; Choi, H.; Han, H.N. Tailoring hetero-grained austenite via a cyclic thermomechanical process for achieving ultrahigh strength-ductility in medium-Mn steel. *Scr. Mater.* **2022**, *217*, 114767. [CrossRef]
13. Li, J.; Xu, Y.; Lu, B.; Yu, Y.; Jing, Y.; Sun, W. Improvement of strength-ductility combination in ultra-high-strength medium-Mn Q&P steel by tailoring the characteristics of martensite/retained austenite constituents. *J. Mater. Res. Technol.* **2022**, *18*, 352–369. [CrossRef]
14. Mohammed, B.; Park, T.; Pourboghrat, F.; Hu, J.; Esmaeilpour, R.; Abu-Farha, F. Multiscale crystal plasticity modeling of multiphase advanced high strength steel. *Int. J. Solids Struct.* **2018**, *151*, 57–75. [CrossRef]
15. Park, T.; Kim, H.; Ryu, I.; Pourboghrat, F.; Esmaeilpour, R. A multi-scale modelling of 3rd generation advanced high strength steels to account for anisotropic evolution of yield surface and plastic potential. *J. Phys. Conf. Ser.* **2018**, *1063*, 012019. [CrossRef]
16. Park, T.; Hector, L.G.; Hu, X.; Abu-Farha, F.; Fellinger, M.R.; Kim, H.; Esmaeilpour, R.; Pourboghrat, F. Crystal plasticity modeling of 3rd generation multi-phase AHSS with martensitic transformation. *Int. J. Plast.* **2019**, *120*, 1–46. [CrossRef]
17. He, B.B.; Luo, H.W.; Huang, M.X. Experimental investigation on a novel medium Mn steel combining transformation-induced plasticity and twinning-induced plasticity effects. *Int. J. Plast.* **2016**, *78*, 173–186. [CrossRef]
18. Kalsar, R.; Sanamar, S.; Schell, N.; Brokmeier, H.G.; Saha, R.; Ghosh, P.; Suwas, S. In-situ study of tensile deformation behaviour of medium Mn TWIP/TRIP steel using synchrotron radiation. *Mater. Sci. Eng. A* **2022**, *857*, 144013. [CrossRef]
19. Nimaga, O.G.; He, B.B.; Cheng, G.J.; Yen, H.W.; Huang, M.X. Revealing orientation-dependent martensitic transformation in a medium Mn steel by micropillar compression. *Int. J. Plast.* **2019**, *123*, 165–177. [CrossRef]
20. Zhang, M.; Chen, H.; Wang, Y.; Wang, S.; Li, R.; Li, S.; Wang, Y.D. Deformation-induced martensitic transformation kinetics and correlative micromechanical behavior of medium-Mn transformation-induced plasticity steel. *J. Mater. Sci. Technol.* **2019**, *35*, 1779–1786. [CrossRef]
21. Soleimani, M.; Kalhor, A.; Mirzadeh, H. Transformation-induced plasticity (TRIP) in advanced steels: A review. *Mater. Sci. Eng. A* **2020**, *795*, 140023. [CrossRef]
22. Lee, Y.K.; Han, J. Current opinion in medium manganese steel. *Mater. Sci. Technol.* **2015**, *31*, 843–856. [CrossRef]
23. Kwok, T.W.J.; Dye, D. A review of the processing, microstructure and property relationships in medium Mn steels. *Int. Mater. Rev.* **2023**, *68*, 1098–1134. [CrossRef]
24. Liu, M.; Zhou, J.; Zhang, J.; Song, C.; Zhai, Q. Ultra-high strength medium-Mn lightweight steel by dislocation slip band refinement and suppressed intergranular κ-carbide with Cr addition. *Mater. Charact.* **2022**, *190*, 112042. [CrossRef]
25. Roters, F.; Eisenlohr, P.; Hantcherli, L.; Tjahjanto, D.D.; Bieler, T.R.; Raabe, D. Overview of constitutive laws, kinematics, homogenization and multiscale methods in crystal plasticity finite-element modeling: Theory, experiments, applications. *Acta Mater.* **2010**, *58*, 1152–1211. [CrossRef]
26. Hu, P.; Liu, Y.; Zhu, Y.; Ying, L. Crystal plasticity extended models based on thermal mechanism and damage functions: Application to multiscale modeling of aluminum alloy tensile behavior. *Int. J. Plast.* **2016**, *86*, 1–25. [CrossRef]
27. Zhang, X.; Liu, Y.; Oskay, C. Multiscale Reduced-Order Modeling of a Titanium Skin Panel Subjected to Thermomechanical Loading. *arXiv* **2021**, .. [CrossRef]
28. Williams, J.C.; Baggerly, R.G.; Paton, N.E. Deformation behavior of HCP Ti-Al alloy single crystals. *Metall. Mater. Trans. A* **2002**, *33*, 837–850. [CrossRef]
29. Ozturk, D.; Shahba, A.; Ghosh, S. Crystal plasticity FE study of the effect of thermo-mechanical loading on fatigue crack nucleation in titanium alloys. *Fatigue Fract. Eng. Mater. Struct.* **2016**, *39*, 752–769. [CrossRef]

30. Venkataramani, G.; Kirane, K.; Ghosh, S. Microstructural parameters affecting creep induced load shedding in Ti-6242 by a size dependent crystal plasticity FE model. *Int. J. Plast.* **2008**, *24*, 428–454. [CrossRef]
31. Oppedal, A.L.; El Kadiri, H.; Tomé, C.N.; Kaschner, G.C.; Vogel, S.C.; Baird, J.C.; Horstemeyer, M.F. Effect of dislocation transmutation on modeling hardening mechanisms by twinning in magnesium. *Int. J. Plast.* **2012**, *30–31*, 41–61. [CrossRef]
32. Beyerlein, I.J.; Tomé, C.N. A dislocation-based constitutive law for pure Zr including temperature effects. *Int. J. Plast.* **2008**, *24*, 867–895. [CrossRef]
33. Xia, D.; Oskay, C. Reduced order mathematical homogenization method for polycrystalline microstructure with microstructurally small cracks. *Int. J. Numer. Methods Eng.* **2023**, *124*, 3166–3190. [CrossRef]
34. Madec, R.; Devincre, B.; Kubin, L.P. From Dislocation Junctions to Forest Hardening. *Phys. Rev. Lett.* **2002**, *89*, 255508. [CrossRef] [PubMed]
35. Liu, Y.; Zhang, X.; Oskay, C. A comparative study on fatigue indicator parameters for near-α titanium alloys. *Fatigue Fract. Eng. Mater. Struct.* **2023**, *46*, 271–294. [CrossRef]
36. Kocks, U.F.; As, A.; Mf, A. *Thermodynamics and Kinetics of Slip*; Pergamon Press: Oxford, UK, 1975.
37. Williams, W.S.; Schaal, R.D. Elastic deformation, plastic flow, and dislocations in single crystals of titanium carbide. *J. Appl. Phys.* **1962**, *33*, 955–962. [CrossRef]
38. Appel, F.; Lorenz, U.; Oehring, M.; Sparka, U.; Wagner, R. Thermally activated deformation mechanisms in micro-alloyed two-phase titanium amminide alloys. *Mater. Sci. Eng. A* **1997**, *233*, 1–14. [CrossRef]
39. Cuitino, A.M.; Ortiz, M. Computational modelling of single crystals. *Model. Simul. Mater. Sci. Eng.* **1993**, *1*, 225. [CrossRef]
40. Kalidindi, S.R.; Bronkhorst, C.A.; Anand, L. Crystallographic texture evolution in bulk deformation processing of FCC metals. *J. Mech. Phys. Solids* **1992**, *40*, 537–569. [CrossRef]
41. Groeber, M.A.; Jackson, M.A. DREAM.3D: A Digital Representation Environment for the Analysis of Microstructure in 3D. *Integr. Mater. Manuf. Innov.* **2014**, *3*, 56–72. [CrossRef]
42. Zhang, X.; Oskay, C. Eigenstrain based reduced order homogenization for polycrystalline materials. *Comput. Methods Appl. Mech. Eng.* **2015**, *297*, 408–436. [CrossRef]
43. Zhang, X.; Oskay, C. Polycrystal plasticity modeling of nickel-based superalloy IN 617 subjected to cyclic loading at high temperature. *Model. Simul. Mater. Sci. Eng.* **2016**, *24*, 055009. [CrossRef]
44. Nasirov, A.; Zhang, X.; Wagner, D.; Yeratapally, S.R.; Oskay, C. Achieving high efficiency in reduced order modeling for large scale polycrystal plasticity simulations. *Finite Elem. Anal. Des.* **2024**, *228*, 104053. [CrossRef]
45. Tian, Y.; Lin, S.; Ko, J.Y.P.; Lienert, U.; Borgenstam, A.; Hedström, P. Micromechanics and microstructure evolution during in situ uniaxial tensile loading of TRIP-assisted duplex stainless steels. *Mater. Sci. Eng. A* **2018**, *734*, 281–290. [CrossRef]
46. Baghdadchi, A.; Hosseini, V.A.; Karlsson, L. Identification and quantification of martensite in ferritic-austenitic stainless steels and welds. *J. Mater. Res. Technol.* **2021**, *15*, 3610–3621. [CrossRef]
47. Cai, Z.H.; Zhang, D.L.; Ma, L.F.; Ding, H.; Feng, Y.; Hu, J.; Misra, R.D.K. Competing deformation mechanisms in an austenite-ferrite medium-Mn steel at different strain rates. *Mater. Sci. Eng. A* **2021**, *818*, 141357. [CrossRef]
48. Benzing, J.T.; Luecke, W.E.; Mates, S.P.; Ponge, D.; Raabe, D.; Wittig, J.E. Intercritical annealing to achieve a positive strain-rate sensitivity of mechanical properties and suppression of macroscopic plastic instabilities in multi-phase medium-Mn steels. *Mater. Sci. Eng. A* **2021**, *803*, 140469. [CrossRef] [PubMed]
49. Lu, Y.; Yu, H.; Sisson, R.D. The effect of carbon content on the c/a ratio of as-quenched martensite in Fe-C alloys. *Mater. Sci. Eng. A* **2017**, *700*, 592–597. [CrossRef]
50. Cheong, K.S.; Busso, E.P. Effects of lattice misorientations on strain heterogeneities in FCC polycrystals. *J. Mech. Phys. Solids* **2006**, *54*, 671–689. [CrossRef]
51. Chen, W.F.; Han, D.J.; Han, D.J. *Plasticity for Structural Engineers*; J. Ross Publishing: Plantation, FL, US, 2007.
52. McDowell, D.L.; Dunne, F.P.E. Microstructure-sensitive computational modeling of fatigue crack formation. *Int. J. Fatigue* **2010**, *32*, 1521–1542. [CrossRef]
53. Dunne, F.P.E.; Wilkinson, A.J.; Allen, R. Experimental and computational studies of low cycle fatigue crack nucleation in a polycrystal. *Int. J. Plast.* **2007**, *23*, 273–295. [CrossRef]

crystals

MDPI

Article

Effects of Stress State, Crack—γ/γ' Phase Interface Relative Locations and Orientations on the Deformation and Crack Propagation Behaviors of the Ni-Based Superalloy—A Molecular Dynamics Study

Xinmao Qin [1,2,3,4], Yilong Liang [1,3,4,*] and Jiabao Gu [1,3,4]

1 College of Materials Science and Metallurgical Engineering, Guizhou University, Guiyang 550025, China; qxm200711@126.com (X.Q.); 15904685778@163.com (J.G.)
2 School of Electronic and Information Engineering, Anshun University, Anshun 561000, China
3 Guizhou Key Laboratory for Mechanical Behavior and Microstructure of Materials, Guiyang 550025, China
4 National & Local Joint Engineering Laboratory for High-Performance Metal Structure Material and Advanced Manufacturing Technology, Guiyang 550025, China
* Correspondence: ylliang@gzu.edu.cn

Abstract: In this study, we systematically investigate the influence of stress states, relative locations, and orientations of crack—γ/γ' phase interfaces on the deformation and crack propagation behaviors of the Ni-based superalloy through molecular dynamics simulations. The stress state with high stress triaxiality will impede the plastic deformation process of the system, thereby promoting brittle crack propagation within the system. But the stress state of low stress triaxiality results in obvious plastic deformation and plastic crack propagation behaviors of the system. The deformation system with cracks located in both the γ and γ' phase exhibits the slowest growth rate, regardless of applied stress states. Additionally, the deformation process demonstrates prominent plastic behavior. For the deformation system with cracks perpendicular to the γ/γ' phase interface, the γ/γ' phase interface will hinder the crack propagation. Our research provides interesting observations on deformation and crack propagation behaviors at an atomic level and at a nano-scale which are important for understanding deformation and fracture behaviors at a macroscopic scale for the Ni-based superalloy.

Keywords: stress state; crack location; crack orientation; crack propagation; Ni-based superalloy

1. Introduction

Due to its exceptional elevated-temperature strength, oxidation resistance, and corrosion resistance, the Ni-based superalloy has been extensively utilized as elevated-temperature components in aircraft engines [1–4], including turbine blades, turbine discs, and combustors. Moreover, it has also found application as high-strength fasteners in wind-driven generators [5]. These critical components inevitably experience damage and fracture during service. Component fracture is a crucial concern that directly impacts integrity and safety [6]. Therefore, understanding the fracture process of materials is essential for enhancing component safety and reliability. At the macroscopic scale, component fractures primarily occur due to corrosion, fatigue, and creep mechanisms, among others. However, at smaller scales, such as the atomic-level crack-scale, small crack growth up to meso-scale length crack growth of materials contributes to the overall fracture process of components. Henceforth, studying atomic-scale crack initiation and propagation in materials becomes imperative for comprehending the fundamental mechanism behind macroscopic scale.

The crack propagation and fracture behaviors of the material have been successfully observed at the atomic scale through the molecular dynamics simulation (MD) method [7–10]. The microstructure evolution of the crack growth and the fracture process is recorded by tracking the motion of the atoms. Wu et al. [11] investigated the influence of temperature on the microstructural evolution and stress distribution during crack propagation in single crystal nickel. They observed that, at elevated temperatures, crack growth was influenced by the formation of voids and slip bands ahead of the crack tip. Zhang et al. [12] and Chang et al. [13] employed molecular dynamics simulations to elucidate the interactions between cracks, dislocations, and microtwins for various lattice orientations. Their findings revealed that the lattice orientation significantly affects crack growth behavior. The influences of grain boundary on the crack growth process in Al and Ni bicrystals were investigated by Fang [14] and Zhang [15], respectively.

The Ni-based superalloy, such as Inconel 718, is primarily composed of the γ base (Ni), γ' (Ni3Al), and γ'' (Ni3Nb) phase [16–18]. The MD method has also successfully investigated the impact of the γ/γ' interface on crack propagation in the Ni-based superalloy. Ma and his colleagues [19] studied the influence of the γ/γ' phase interface on fatigue crack propagation and discovered that it hinders crack growth. Furthermore, the influence of the γ/γ' phase interface on the void growth and microstructure evolution has been analyzed, and the research results are as follows. Jing et al. [20] found that by creating minor diameter holes at the γ/γ' phase interface, the yield strength of the materials can be improved. However, Liu et al. [21] demonstrated that the presence of holes at this interface reduces both yield strength, elastic modulus, and the plastic deformation capacity of materials. Cui′s study [22] shows that holes in the γ/γ' interface promote dislocation slip from the γ phase to γ' phase, thereby reducing mechanical strength. Nevertheless, there has not been any systematic work so far to reveal how stress states, crack locations, and crack orientations affect deformation and crack propagation behaviors as well as mechanisms for Ni-based superalloys.

In this study, we employed the MD simulation method to systematically investigate the influence of stress states, relative locations, and orientations of crack—γ/γ' phase interfaces on deformation and crack growth behaviors in Ni-based superalloys. Our research provides valuable insights into the nano-scale, which are crucial for understanding macroscopic deformation and fracture behaviors in Ni-based superalloys. For example, after the ultrasonic surface rolling modification (USRM), Peng [23] found that the USRM led to the crack initiation position being transferred to about 800 μm from the specimen surface (the crack initiation position of the specimen without USRM is located on the specimen surface). This migration behavior significantly reduces the tension–tension fatigue limit in small-diameter specimens made of the Inconel 718 alloy. Our findings offer a microscopic description of internal fatigue crack propagation in Ni-based superalloys under realistic operating conditions.

2. Simulation Method and Details

2.1. Simulation Method and Model

In this work, the LAMMPS [24] (large-scale atomic/molecular massively parallel simulator) software was utilized to investigate the deformation and crack growth behaviors in a Ni-based superalloy. The simulation model consisted of two phases: the γ phase (Ni) and the γ' phase (Ni_3Al). The atomic arrangement of the two phases is shown in Figure 1f. The dimensions of the model were 51.4 nm, 25.7 nm, and 25.7 nm along the X ([100]), Y ([010]), and Z ([001]) directions, respectively. It included approximately three million atoms. To eliminate lattice mismatch stress between the γ phase and γ' phase, their dimensions in the simulation model were chosen as $na_\gamma \approx ma_{\gamma'}$ (where n and m are positive integers). The lattice constant for γ (Ni) and γ' (Ni_3Al) are 3.524 Å and 3.573 Å, respectively, which indicates that n = 73 and m = 72. Consequently, the size of γ (Ni) was set as $73a_\gamma \times 73a_\gamma \times 73a_\gamma$ while the size of γ' (Ni_3Al) was set as $72a_{\gamma'} \times 72a_{\gamma'} \times 72a_{\gamma'}$. The interface between these two phases occurred at plane (100). By selectively removing some atoms from specific regions within the simulated system, models with specified cracks were created. As depicted in Figure 1, four types of crack locations were considered: within the γ phase (Figure 1a); within γ' phase (Figure 1b); at the interface between γ and γ' phase (Figure 1c); and within both the γ phase and γ' phase (Figure 1d). Additionally, to investigate the influence of crack orientations on the deformation and crack growth behaviors, two types of orientation relationships were taken into account: the long axis direction of crack was parallel to the interface between γ and γ' phase (Figure 1c), while it was perpendicular to this interface (Figure 1e).

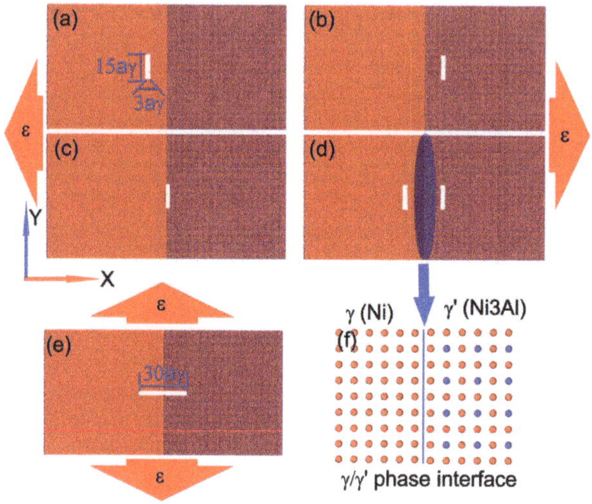

Figure 1. The MD model of Ni-based superalloy with crack.

The three-dimensional periodic boundary conditions are used in the simulation model. Prior to deformation, a relaxation process of 10 ps is conducted at a temperature of 300 K and a pressure of 0 bar using an isothermal–isobaric ensemble (NPT) [25–27]. Subsequently, cyclic deformation is induced by applying strain. The NVE ensemble is employed for the cyclic deformation process with increasing strain amplitude. The strain rate in the X direction remains constant at 1×10^{-3} ps^{-1} for all models. The strain ratio ($R = \varepsilon_{min}/\varepsilon_{max} = 0.5$) governs the cyclic deformation process, as depicted in Figure 2. Each cycle consists of loading and unloading stages lasting for 10 ps and 5 ps, respectively. The maximum cyclic numbers of all systems is set to N = 40. It should be noted that three calculations were performed independently for each model to ensure result repeatability. Four different stress states are obtained by applying distinct deformation conditions (Figure 3), which are detailed in Table 1 along with their specific parameters (NO.1, NO.2, NO.3, and NO.4).

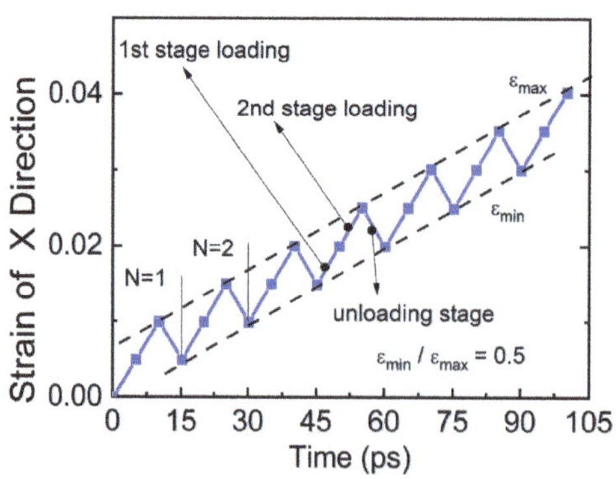

Figure 2. The strain–time relationship in the cyclic deformation with increasing strain amplitude.

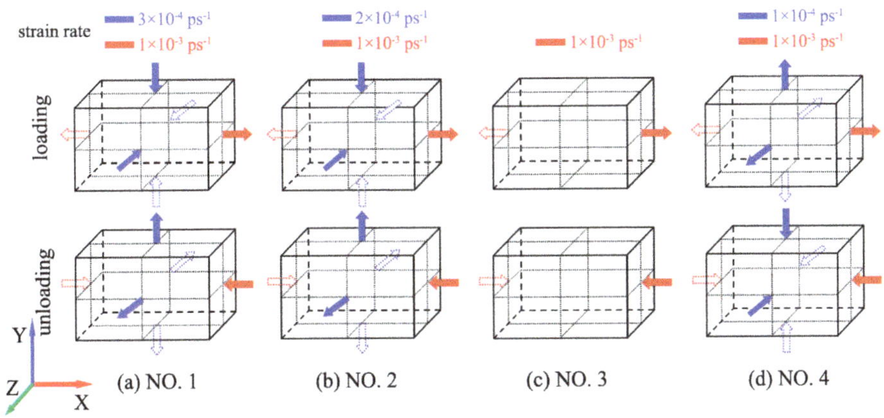

Figure 3. The cyclic loading–unloading schematic diagram under different stress states: (**a**) NO. 1, (**b**) NO. 2, (**c**) NO. 3, and (**d**) NO. 4.

Table 1. The specific parameters (strain rate) of four types of different stress states.

Stress States	X Direction		Y and Z Direction
NO. 1	loading	$+1 \times 10^{-3}$ ps^{-1}	-3×10^{-4} ps^{-1}
	unloading	-1×10^{-3} ps^{-1}	$+3 \times 10^{-4}$ ps^{-1}
NO. 2	loading	$+1 \times 10^{-3}$ ps^{-1}	-2×10^{-4} ps^{-1}
	unloading	-1×10^{-3} ps^{-1}	$+2 \times 10^{-4}$ ps^{-1}
NO. 3	loading	$+1 \times 10^{-3}$ ps^{-1}	0
	unloading	-1×10^{-3} ps^{-1}	0
NO. 4	loading	$+1 \times 10^{-3}$ ps^{-1}	$+1 \times 10^{-4}$ ps^{-1}
	unloading	-1×10^{-3} ps^{-1}	-1×10^{-4} ps^{-1}

2.2. The Interatomic Pairwise Interaction Potentials

In this study, the interatomic pairwise interaction potential of the γ(Ni)/γ'(Ni_3Al) system was described with an EAM/alloy potential [28]. The EAM/alloy potential provided the total energy of system atoms as:

$$E_i = F_\alpha\left(\sum_{j\neq i}\rho_\beta(r_{ij})\right) + 1/2\cdot\sum_{j\neq i}\varnothing_{\alpha\beta}(r_{ij}) \quad (1)$$

where F is the embedding energy which is a function of the atomic electron density ρ (rho); $\varnothing$ (phi) is a pair potential interaction; a and b are the element types of atoms I and J. The relevant parameters of the EAM/alloy potential are listed in Table 2.

Table 2. The relevant parameters of the Ni-Al interatomic EAM/alloy potential.

Parameter	Value
N_{rho}	10,000
d_{rho}	$0.5013228884 \times 10^{-3}$
N_r	10,000
d_r	0.62872×10^{-3}
cutoff	6.28721

where N_{rho} and N_r are the number of tabulated values in the subsequent arrays; d_{rho} and d_r are the spacing in density and distance space for the values in those arrays.

2.3. The Analysis Method of the Microstructure and the Stress Definition

To analyze the deformation and crack propagation behaviors of these systems, the microstructure evolution of the model are analyzed by common neighbor analysis (CNA) [29,30] and the dislocation extraction algorithm (DXA) of the OVITO [31,32] software (the open visualization tool).

The elements of the atomic-level stress tensor are calculated in this work, which includes the atomic stress ($\sigma_{\alpha\beta}(i)$), the hydrostatic stress (σ_m) of the entire system, the effective von Mises stress (σ_e) of the entire system, and the stress triaxiality (Tr) of the entire system [33,34]. The atomic stress ($\sigma^{\alpha\beta}(i)$) is calculated by:

$$\sigma_{\alpha\beta(i)} = -1/\Omega_0 \cdot \left[1/2 \cdot \sum_j F_{ij}^{\alpha} r_{ij}^{\beta} + M_i v_i^{\alpha} v_i^{\beta}\right] \tag{2}$$

where a and b are the Cartesian components; Ω_0 is the volume of the atoms; F_{ij} is the force exerted by atom j on atom i; r_{ij} is the vector component form of the relative position from atom j to atom i; M_i is the mass of atom i; and v_i is the velocity of atom i. The hydrostatic stress (σ_m) of the entire system is calculated by:

$$\sigma_m = 1/3 \cdot (\sigma_x + \sigma_y + \sigma_z) \tag{3}$$

where σ_x, σ_y, and σ_z are the stresses averaged over the entire system in the X, Y, and Z directions, respectively. The effective von Mises stress (σ_e) of the entire system is calculated by:

$$\sigma_e = \left\{1/2 \cdot \left[(\sigma_x - \sigma_y)^2 + (\sigma_x - \sigma_z)^2 + (\sigma_y - \sigma_z)^2\right] + 6\left(\sigma_{xy}^2 + \sigma_{yz}^2 + \sigma_{xz}^2\right)\right\}^{1/2} \tag{4}$$

Then, the stress triaxiality of the entire system is given by:

$$Tr = \sigma_m / \sigma_e \tag{5}$$

3. Results and Discussions

3.1. The Effect of Stress States on the Deformation and Crack Propagation Behaviors

3.1.1. The Deformation Behaviors

The initial stress triaxiality, stress-strain relationship, and characteristic strain of the deformation system under various stress states is illustrated in Figure 4. The characteristic strain of the deformation system encompasses the total strain (ε_t), elastic strain (ε_e), and accumulated plastic strain ($\varepsilon_p = \varepsilon_t - \varepsilon_e$). It is important to note that the definition of total strain (ε_t) is divided into two scenarios: (i) when the system experiences complete fracture during cyclic deformation, it is defined as the strain at which fracture occurs; (ii) when the system does not undergo complete fracture, it is defined as the total strain value after 40 cycles.

In the cyclic deformation process, as the stress state of the deformation system changes from NO. 1 to NO. 4, there is an increase in the initial stress triaxiality of the system from 0.25 to 0.95. For low initial stress triaxiality (the stress state of NO. 1 and NO. 2), the cyclic tensile stress gradually decreases after reaching its maximum value, with plastic deformation dominating throughout the entire cyclic deformation process. However, for high initial stress triaxiality (the stress state of NO. 4), the cyclic tensile stress rapidly drops close to zero GPa after reaching its maximum value, and plastic deformation only occurs briefly before reaching this peak (Figure 4b; shaded area). We further calculated the total cyclic strain (ε_t), elastic strain (ε_e), and cumulative plastic strain (ε_p), as shown in Figure 4c. It is evident that under different stress states (NO.1, NO. 2, NO. 3, and NO. 4), accumulated plastic strains are measured at approximately 19.34%, 18.13%, 10.14%, and 2.58%, respectively. With increasing initial stress triaxiality from NO. 1 to NO. 4, the initial stress triaxiality increases; thus, the deformation behavior transitions from typical plastic behavior to brittle behavior.

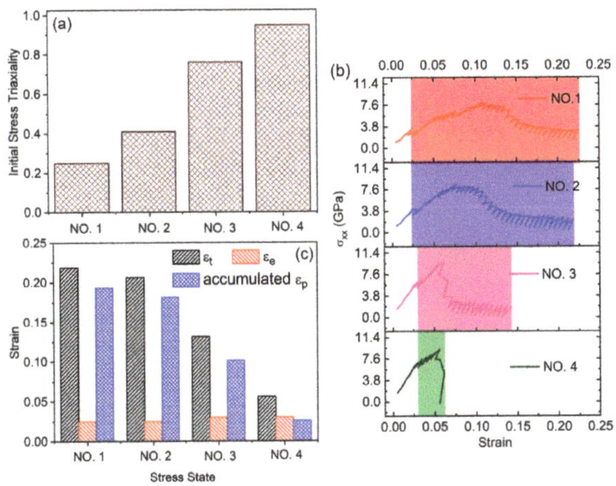

Figure 4. The relationship of the stress and strain of the deformation system under various stress states, where (**a**) is the initial stress triaxiality variation, (**b**) is the stress–strain curve, and (**c**) is the characteristic strain variation.

Under conditions of low stress triaxiality, the normal stress experienced by the atoms in the system is relatively small, while the shear stress is comparatively large. This enhances the likelihood of dislocation slip within the system, resulting in pronounced plastic deformation behavior. Conversely, under high stress triaxiality conditions, the normal stress exerted on the atoms becomes significant while the shear stress diminishes. Consequently, dislocations encounter greater difficulty in movement and ultimately exhibit conspicuous brittle behavior [35–38].

3.1.2. The Crack Length—Cyclic Numbers Relationship

According to the relationship between crack length and cyclic numbers shown in Figure 5, the crack growth can be divided into three stages: stage I—crack growth incubation stage (elliptic shadow region), stage II—crack rapid growth stage (inverted rectangular shadow region), and stage III—crack stable growth stage (square shadow region). It is evident that the cyclic numbers for each stage of crack growth in the deformation system vary significantly with stress states. The stress states have a remarkable influence on the crack propagation process. As the initial stress triaxiality of the system increases from 0.25 to 0.95, there is a reduction in the required cyclic numbers for each stage of crack growth, leading to an obvious acceleration in the rate of crack propagation.

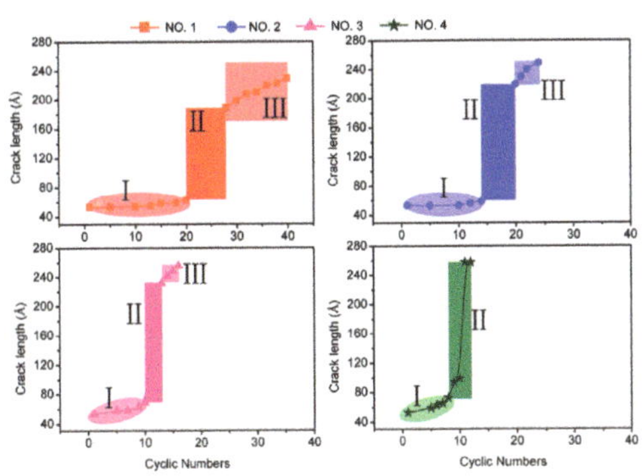

Figure 5. The crack length versus the cyclic numbers curve of the deformation system under various stress states. The symbols I, II, and III represent different stages of crack growth.

3.1.3. The Microstructure Evolution and Dislocations Density Analysis

We employed slice processing to acquire the microstructure of the regime exhibiting the maximum crack length, as shown in Figure 6(a1,b1,c1,d1). For the system with stress state NO. 1, from Figure 6(a1–a6), it can be observed that numerous stacking faults (red atomic region—HCP structure) emerge during the cyclic deformation process, indicating that crack propagation is accompanied by continuous emission and slip of dislocations. The emission of dislocations from the crack tip and their subsequent slip on its slip plane facilitate bond breaking within the crack tip region, ultimately leading to continuous plastic propagation of cracks. The density of dislocations in the system monotonically increases with each cycle number, further confirming that plastic deformation dominates throughout most of the cyclic deformation process, as depicted in Figure 7. However, when subjected to high initial stress triaxiality, crack growth occurs in three stages: nanopore formation, nanopore growth, and merging of nanopores with main cracks (Figure 6d). This type of crack growth exhibits a significantly higher propagation rate. For stress states NO. 2, NO. 3, and NO. 4, the density of dislocations in the deformation system decreases after reaching its peak value due to complete fracture occurring in certain regions within this system. In addition, we also analyzed the fracture morphology after crack propagation (Figure 6(a7,b7,c7,d7)). It can be observed that under stress state NO. 1, after undergoing 40 cycles, the fracture morphology becomes rough and uneven for this deformation system. However, under stress state NO. 4, the entire system fractures completely after only 11 cycles, and exhibits a relatively flat fracture morphology.

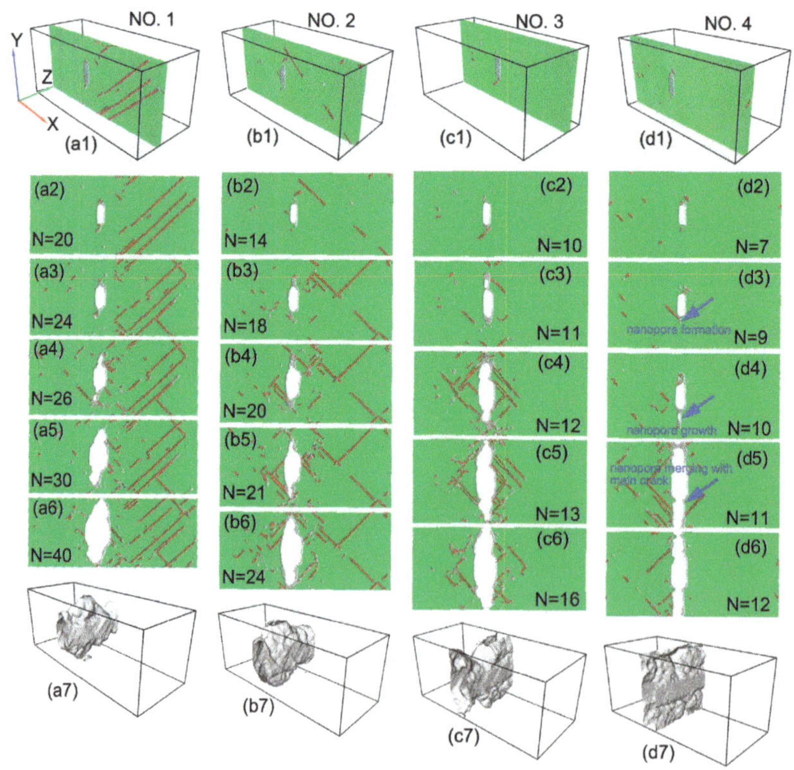

Figure 6. The microstructure evolution of the deformation system under different stress states.

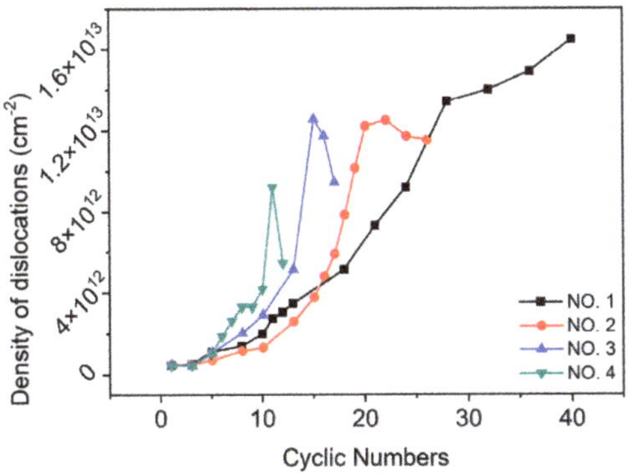

Figure 7. The dislocations density versus the cyclic numbers curve of the deformation system under various stress states.

3.2. The Effect of Crack Locations on the Deformation and Crack Propagation Behaviors

3.2.1. The Deformation Behaviors

Figure 8a shows the initial stress triaxiality of the deformation system under stress state of NO. 3 with different crack locations. It is found that the deformation system with the crack only located in the γ phase has the minimum initial stress triaxiality (T_r = 0.76). The stress triaxiality of the deformation system with other crack locations (such as only located in the γ' phase, γ/γ' phase interface, and both located in the γ and γ' phases) is about 0.84, which indicates that the crack locations have a minor effect on the stress response of the deformation system.

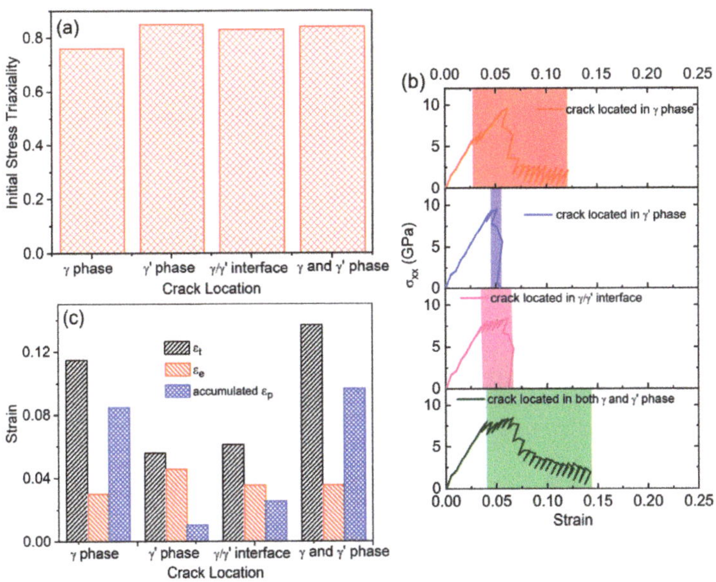

Figure 8. The relationship of the stress and strain of the deformation system under various crack locations, where (**a**) is the initial stress triaxiality variation, (**b**) is the stress-strain curve, and (**c**) is the characteristic strain variation.

As shown in Figure 8b, for the system with the crack only located at the γ phase, when the cyclic deformation strain exceeds about 2.5%, the system has entered the plastic deformation stage. Thereafter, plastic deformation continues until the system completely fractured. For the deformation system where the crack is only located in the γ' phase, an extremely short plastic deformation process occurs before the cyclic tensile stress reaches its maximum value (blue curve in Figure 8b). When the crack is located at the γ/γ' phase interface and the cracks are located symmetrically in the γ and γ' phases, the cyclic tensile stress–strain curve of the system is obviously a zigzag shape, which indicates these two systems have a clear plastic deformation process.

From Figure 8c, it is evident that when the crack solely exists in the γ' phase, the system exhibits pronounced brittle behavior during cyclic deformation. The accumulated plastic strain is approximately 1%, and plastic strain only occurs briefly before crack propagation. In contrast, for a system where the crack resides at the interface between the γ/γ' phases, plastic deformation behavior becomes more apparent compared to a system with cracks solely located in the γ' phase, resulting in a slightly larger accumulated plastic strain (2.55%). Both of these crack locations demonstrate obvious brittle behavior. However, when cracks are exclusively present in the γ phase or simultaneously exist in both the γ and γ' phases, significant plastic behavior is observed with respective accumulated plastic strains of 8.48% and 9.67%. These findings indicate that relative locations of cracks with respect to the γ/γ' phase interface have a notable influence on system deformation behavior; specifically, exhibiting brittle behavior when cracks are situated within the γ' phase or at its interface while demonstrating plastic behaviors when cracks reside solely within the γ. Notably, symmetrically positioned cracks within both γ and γ' phases result in optimal plastic deformation ability.

3.2.2. The Crack Length—Cyclic Numbers Relationship

The curves in Figure 9 illustrate the relationship between crack length and cyclic numbers for a deformation system with varying crack locations. For the systems where the crack is located only at the γ phase, the γ/γ' phase interface, and cracks are located in both γ and γ' phases, the crack growth incubation stage is 10 cyclic numbers. However, there are significant differences in the number of cyclic numbers required for rapid growth and stable growth stages. As we observe changes in crack locations from both in γ and γ' phase, only in γ phase, γ/γ' phase interface to γ' phase, there is a gradual reduction in cyclic numbers necessary for crack growth and the crack growth rate increases. Notably, when two cracks are symmetrically distributed within the system, the slowest rate of crack propagation is exhibited.

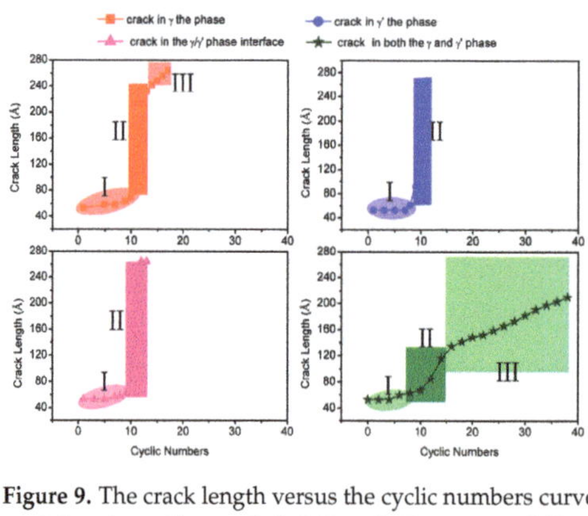

Figure 9. The crack length versus the cyclic numbers curve of the deformation system under various crack locations. The symbols I, II, and III represent different stages of crack growth.

3.2.3. The Microstructure Evolution and Dislocation Density Analysis

The microstructure evolution of crack propagation in the deformation system, where cracks are present in both the γ and γ' phases, is illustrated in Figure 10. It can be observed that the symmetric distribution of two cracks with respect to the γ/γ' phase interface leads to their growth following a typical plastic behavior. As depicted in Figure 10(a3–a5), during the initial stage of crack propagation, dislocation emissions passivate the crack tip, resulting in slow crack propagation due to a continuous dislocation motion. With an increase in cyclic numbers, such as when N = 16, dislocations emitted by both cracks converge at the γ/γ' phase interface and interact with each other (indicated by the red arrow in Figure 10(a6)), leading to coalescence of the two cracks into a wider crack along the X direction. Subsequently, this broader crack gradually propagates forward at a moderate rate.

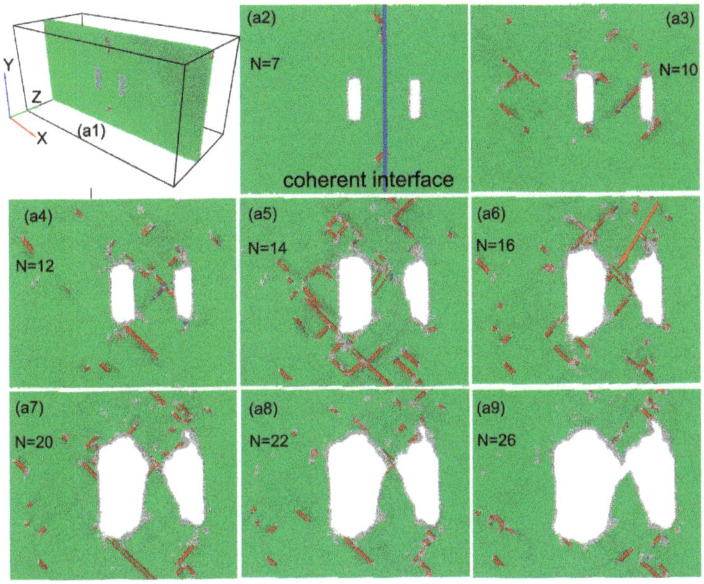

Figure 10. The microstructure evolution of the deformation system with crack located in both the γ and γ' phase. The red arrow on the subfigure (**a6**) indicates a dislocation slip band between the two cracks.

The microstructure evolution and stress distribution of the deformation system with a crack solely located at the γ' phase are depicted in Figure 11. At N = 8, the crack initiates propagation and continues for only two cycles before complete fracture of the system occurs. This can be attributed to the limited plastic deformation capability during deformation, leading to unrelieved stress concentrations at the crack tip. Consequently, rapid crack propagation takes place without significant plastic deformation behavior, resulting in swift and smooth fracture of the system within an extremely short number of cycles (Figure 11(a5)).

The dependence of the dislocation density on the cyclic number is illustrated in Figure 12. It can be observed that when the crack is solely located at the γ phase and when cracks are distributed in both the γ and γ' phases, there is a relatively high maximum dislocation density during cyclic deformation. This can primarily be attributed to the fact that the γ phase serves as the main plastic phase within the system, possessing a accommodating plastic deformation. Conversely, when the crack is exclusively situated at either the γ' phase or at the interface between γ and γ', there exists a relatively low maximum dislocation density throughout cyclic deformation. The main reason is that the γ' phase is a hard and brittle phase with weak plastic deformation capacity, this leads to the reduced production of dislocations emission and plastic deformation.

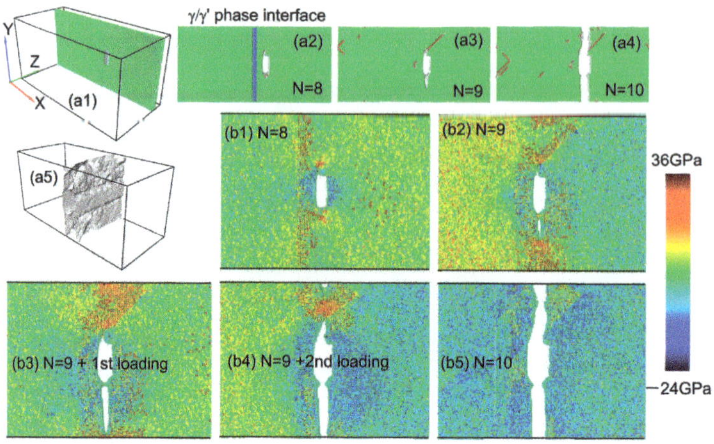

Figure 11. The microstructure evolution and stress distribution of the deformation system with crack located at the γ' phase.

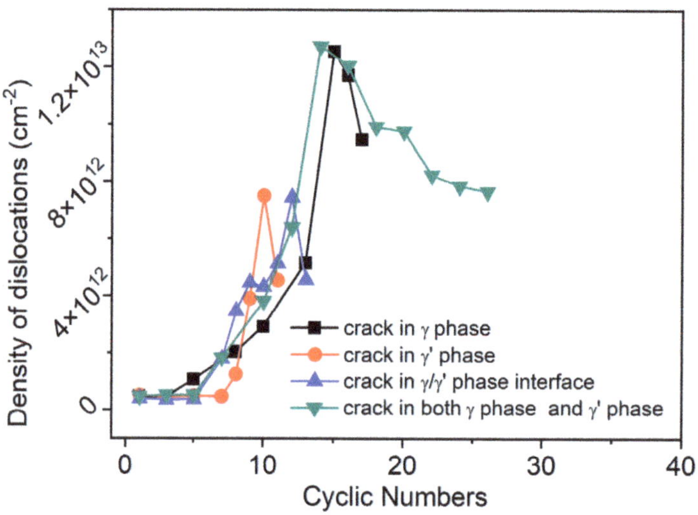

Figure 12. The dislocations density versus the cyclic numbers curve of the deformation system under various crack distributions.

3.3. The Effect of Crack Orientations on the Deformation and Crack Propagation Behaviors

3.3.1. The Deformation Behaviors

Figure 13a demonstrates how relative orientation between crack and phase interface affects initial stress triaxiality. It can be observed that as stress state changes NO. 1 to NO. 4, there is a gradual increase in initial stress triaxiality within this system. Furthermore, regardless of stress state, systems with crack perpendicular to interface exhibit larger initial stress triaxiality compared to those with crack parallel to interface. These findings highlight that relative orientation of crack and interface significantly influences system's stress response, particularly under large stress triaxiality.

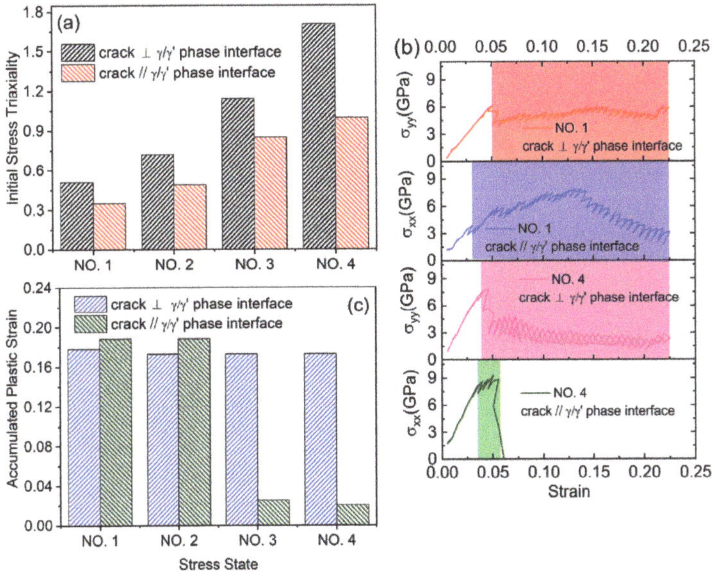

Figure 13. The relationship of the stress and strain of the deformation system under various crack orientations. And (**a**) is the initial stress triaxiality variation, (**b**) is the stress-strain curve, and (**c**) is the characteristic strain variation.

Figure 13b depicts the cyclic tensile stress-strain behavior of the system with varying orientations between the crack and γ/γ' phase interface under NO. 1 and NO. 4 stress states. It is observed that, under NO. 1 stress state, plastic deformation initiates at a strain of approximately 5% for systems with cracks perpendicular to the phase interface, while cyclic tensile stress remains constant throughout plastic deformation. Conversely, for systems with cracks parallel to the phase interface, plastic behavior occurs only at a strain of about 3%, after which cyclic tensile stress gradually decreases following its maximum value. The shaded regions in Figure 13b represent zones of plastic strain within the deformation system.

The accumulated plastic strain of the system with different orientations between the crack and γ/γ' phase interface under stress states NO. 1 to NO. 4 is illustrated in Figure 13c. Minor effects on the accumulative plastic strain of the system during cyclic deformation are observed for stress states NO. 1 and NO. 2, regardless of the relative orientation between the crack and interface. However, significant influences on the accumulated plastic strain are observed at stress states NO. 3 and NO. 4, where systems with a crack perpendicular to the interface exhibit much larger values compared to those with a crack parallel to the interface. Therefore, it can be concluded that systems with a crack perpendicular to the interface demonstrate excellent capacity for plastic deformation.

3.3.2. The Crack Length—Cyclic Numbers Relationship

At the stress state of NO. 1, for systems with crack parallel to the interface, the crack propagation process consists of three typical stages (Figure 14), namely, the crack growth incubation stage (about 25 cyclic numbers), followed by a rapid growth stage (only 3 cyclic numbers), and finally reaching a stable growth stage (12 cyclic numbers). When subjected to stress state NO. 4, there are only two stages in the crack growth process: an incubation stage (around 9 cyclic numbers) and a rapid growth stage (only 2 cyclic numbers). In this particular orientation relationship, the system exhibits an exceptionally fast crack propagation rate, requiring only 11 cyclic numbers for complete fracture. However, during the entire cyclic deformation process of systems with cracks perpendicular to the interface, the crack propagates at a moderate rate from its initial length of 107 Å to reach a length of only 183

Å. Under stress state NO. 4, both the incubation and rapid growth stages for systems with this orientation relationship are very short (only lasting for approximately 7 and 4 cyclic numbers, respectively), while the stable growth stage is significantly prolonged spanning over 29 cyclic numbers.

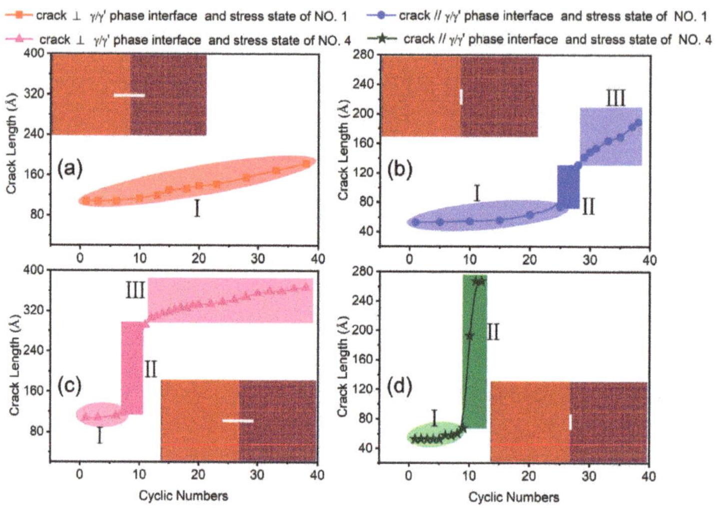

Figure 14. The crack length versus the cyclic numbers curve of the deformation system under various crack orientations. The symbols I, II, and III represent different stages of crack growth.

3.3.3. The Microstructure Evolution and Dislocation Density Analysis

The microstructure evolution of the system with a crack perpendicular to the γ/γ' phase interface under stress states NO. 1 and NO. 4 is depicted in Figure 15. Due to the low stress triaxiality observed in stress state NO. 1, the system exhibits excellent plastic deformation ability, allowing for passivation of the crack tip through continuous dislocation emission. The system experiences secondary strengthening effects as a result of the continuous dislocation slip and dislocation reaction. Interestingly, when the system is deformed under the stress state of NO. 4, the crack propagates only on one side of the γ' phase until it fully traverses through the γ' phase, subsequently meeting with the next γ/γ' phase interface (Figure 15(b5)). Once this transition occurs and the crack growth reaches stability at this new interface, further propagation becomes hindered due to significant stress constraints impeding dislocation emission at the crack tip.

The dislocation density—cyclic numbers curve of the system with different orientations between the crack and γ/γ' phase interface is illustrated in Figure 16. Under the stress state NO. 1, there is little difference in the dislocation density of the final deformation state between systems with different crack—phase interface orientations. This can be attributed to continuous dislocation emission, slip, dislocation interaction, and increment occurring during the deformation process of both systems under this stress state. However, for stress state NO. 4, when reaching the final deformation state, it is observed that the system with a crack perpendicular to the phase interface exhibits a significantly higher dislocation density compared to that with a crack parallel to the phase interface. This phenomenon is due to the fact that when the cyclic numbers are N = 15, crack propagation in the γ' phase has been completed, and the crack tip meets with the next γ/γ' phase interface (Figure 15(b5)), at which the dislocation density of the system reaches the maximum. In the subsequent deformation process, due to the hindering effect of the next γ/γ' phase interface on crack propagation (Figure 15(b6)), the crack hardly continues to propagate, resulting in the dislocations density remaining almost constant during the process. This result is consistent with the research of Xiao et al. [19].

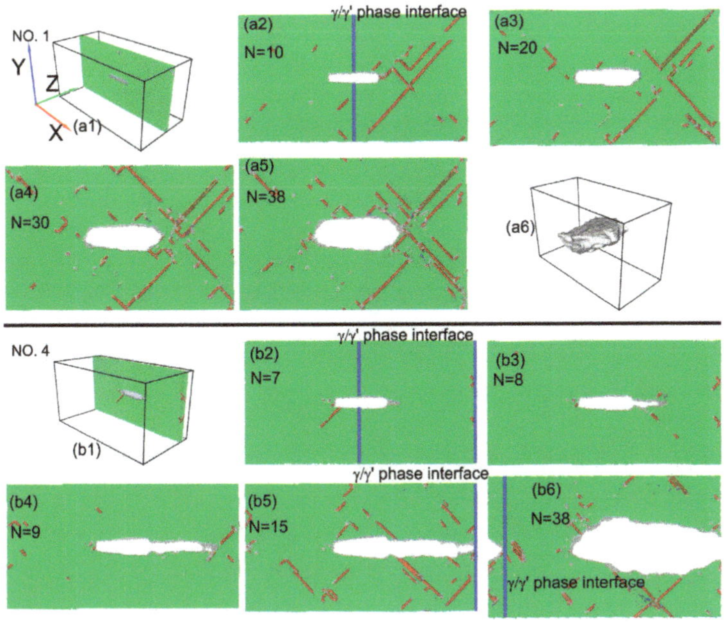

Figure 15. The microstructure evolution of deformation system of the crack ⊥ γ/γ' phase interface under stress state of NO. 1 and NO. 4.

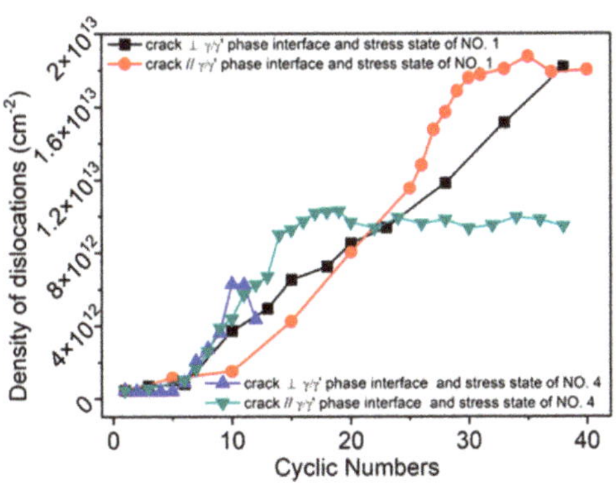

Figure 16. The dislocations density versus the cyclic numbers curve of the deformation system under various crack orientations.

4. Conclusions

The deformation and crack propagation behaviors of the Ni-based superalloy were investigated using MD simulations, with a focus on the influence of stress states, crack—γ/γ' phase interface relative locations, and orientation. The main conclusions were drawn, as follows:

1. With the change in stress state from low stress triaxiality to high stress triaxiality, there is a shift in deformation behavior and crack propagation mode from ductile behavior to brittle behavior.
2. The deformation system with cracks located in both the γ and γ' phase exhibits the slowest growth rate under any applied stress states, while demonstrating obvious plastic behavior during the deformation.

3. For the deformation system with a crack perpendicular to the γ/γ' phase interface, the γ/γ' phase interface will hinder the crack propagation.

Our research provides interesting observations of the deformation and crack propagation behaviors on an atomic level and at nano-scale which are important for understanding deformation and fracture behaviors at a macroscopic scale for the Ni-based superalloy.

Author Contributions: X.Q.: Writing original manuscript; Y.L.: Review and editing manuscript; J.G.: Visualization and analysis. All authors have read and agreed to the published version of the manuscript.

Funding: This work was finally funded by the Engineering Technology Research Center (NO. [2019]5303), the central government guides local science and technology development (NO. [2019]4011) and the Project of the Education Department of Guizhou Province (NO. [2021]315).

Data Availability Statement: All relevant data are within the paper.

Acknowledgments: The authors are grateful for the research support of Yan Wanjun in massively parallel computing.

Conflicts of Interest: The authors declare no conflict of interest.

References

1. Pollock, T.M.; Tin, S. Nickel-Based Superalloys for Advanced Turbine Engines: Chemistry, Microstructure and Properties. *J. Propuls. Power* **2006**, *22*, 361–374. [CrossRef]
2. Miao, J.; Pollock, T.M.; Wayne Jones, J. Crystallographic fatigue crack initiation in nickel-based superalloy René 88DT at elevated temperature. *Acta Mater.* **2009**, *57*, 5964–5974. [CrossRef]
3. Pineau, A.; Antolovich, S.D. High temperature fatigue of nickel-base superalloys—A review with special emphasis on deformation modes and oxidation. *Eng. Fail. Anal.* **2009**, *16*, 2668–2697. [CrossRef]
4. Texier, D.; Stinville, J.-C.; Echlin, M.P.; Pierret, S.; Villechaise, P.; Pollock, T.M.; Cormier, J. Short crack propagation from cracked non-metallic inclusions in a Ni-based polycrystalline superalloy. *Acta Mater.* **2019**, *165*, 241–258. [CrossRef]
5. Andrew, D.L.; Carlson, S.S.; Macha, J.H.; Pilarczyk, R.T. Investigating and interpreting failure analysis of high strength nuts made from nickel-base superalloy. *Eng. Fail. Anal.* **2017**, *74*, 35–53. [CrossRef]
6. Qin, X.; Liang, Y.; Gu, J.; Peng, G. The Effect of Interatomic Potentials on the Nature of Nanohole Propagation in Single-Crystal Nickel: A Molecular Dynamics Simulation Study. *Crystals* **2023**, *13*, 585. [CrossRef]
7. Guin, L.; Raphanel, J.L.; Kysar, J.W. Atomistically derived cohesive zone model of intergranular fracture in polycrystalline graphene. *J. Appl. Phys.* **2016**, *119*, 245107. [CrossRef]
8. Zhang, Y.; Jiang, S.; Zhu, X.; Zhao, Y. Mechanisms of crack propagation in nanoscale single crystal, bicrystal and tricrystal nickels based on molecular dynamics simulation. *Results Phys.* **2017**, *7*, 1722–1733. [CrossRef]
9. Elapolu, M.S.R.; Tabarraei, A. Atomistic Simulation-Based Cohesive Zone Law of Hydrogenated Grain Boundaries of Graphene. *J. Phys. Chem. C* **2020**, *124*, 17308–17319. [CrossRef]
10. Elapolu, M.S.R.; Tabarraei, A. Mechanical and Fracture Properties of Polycrystalline Graphene with Hydrogenated Grain Boundaries. *J. Phys. Chem. C* **2021**, *125*, 11147–11158. [CrossRef]
11. Wu, W.-P.; Yao, Z.-Z. Molecular dynamics simulation of stress distribution and microstructure evolution ahead of a growing crack in single crystal nickel. *Theor. Appl. Fract. Mech.* **2012**, *62*, 67–75. [CrossRef]
12. Zhang, J.; Ghosh, S. Molecular dynamics based study and characterization of deformation mechanisms near a crack in a crystalline material. *J. Mech. Phys. Solids* **2013**, *61*, 1670–1690. [CrossRef]
13. Chang, L.; Kitamura, T.; Zhou, C.-Y. Atomic simulation of the orientation effects on crack tip behavior in titanium single crystal. *Theor. Appl. Fract. Mech.* **2020**, *110*, 102791. [CrossRef]
14. Fang, W.; Xie, H.; Yin, F.; Li, J.; Khan, D.F.; Fang, Q. Molecular dynamics simulation of grain boundary geometry on crack propagation of bi-crystal aluminum. *Mater. Sci. Eng. A* **2016**, *666*, 314–319. [CrossRef]
15. Zhang, Y.; Jiang, S.; Zhu, X.; Zhao, Y. A molecular dynamics study of intercrystalline crack propagation in nano-nickel bicrystal films with (0 1 0) twist boundary. *Eng. Fract. Mech.* **2016**, *168*, 147–159. [CrossRef]
16. Moverare, J.J.; Gustafsson, D. Hold-time effect on the thermo-mechanical fatigue crack growth behaviour of Inconel 718. *Mater. Sci. Eng. A* **2011**, *528*, 8660–8670. [CrossRef]
17. Prasad, K.; Sarkar, R.; Ghosal, P.; Kumar, V. Simultaneous creep–fatigue damage accumulation of forged turbine disc of IN 718 superalloy. *Mater. Sci. Eng. A* **2013**, *572*, 1–7. [CrossRef]
18. Mahobia, G.S.; Paulose, N.; Mannan, S.L.; Sudhakar, R.G.; Chattopadhyay, K.; Santhi Srinivas, N.C.; Singh, V. Effect of hot corrosion on low cycle fatigue behavior of superalloy IN718. *Int. J. Fatigue* **2014**, *59*, 272–281. [CrossRef]
19. Ma, L.; Xiao, S.; Deng, H.; Hu, W. Tensile mechanical properties of Ni-based superalloy of nanophases using molecular dynamics simulation. *Phys. Status Solidi* **2016**, *253*, 726–732. [CrossRef]

20. Shang, J.; Yang, F.; Li, C.; Wei, N.; Tan, X. Size effect on the plastic deformation of pre-void Ni/Ni_3Al interface under uniaxial tension: A molecular dynamics simulation. *Comput. Mater. Sci.* **2018**, *148*, 200–206. [CrossRef]
21. Liu, H.; Wang, X.M.; Liang, H.; Zhao, Z.N.; Li, L.; Yue, Z.F.; Deng, C.H. The effect of void defect on the evolution mechanisms of dislocations and mechanical properties in nickel-based superalloys by molecular dynamics simulation of real γ/γ' structures. *Int. J. Solids Struct.* **2020**, *191–192*, 464–472. [CrossRef]
22. Cui, C.; Gong, X.; Chen, L.; Xu, W.; Chen, L. Atomic-scale investigations on dislocation-precipitate interactions influenced by voids in Ni-based superalloys. *Int. J. Mech. Sci.* **2022**, *216*, 106945. [CrossRef]
23. Peng, X.; Liang, Y.; Qin, X.; Gu, J. The effect of ultrasonic surface rolling process on tension-tension fatigue limit of small diameter specimens of Inconel 718 superalloy. *Int. J. Fatigue* **2022**, *162*, 106964. [CrossRef]
24. Thompson, A.P.; Aktulga, H.M.; Berger, R.; Bolintineanu, D.S.; Brown, W.M.; Crozier, P.S.; in 't Veld, P.J.; Kohlmeyer, A.; Moore, S.G.; Nguyen, T.D.; et al. LAMMPS—A flexible simulation tool for particle-based materials modeling at the atomic, meso, and continuum scales. *Comput. Phys. Commun.* **2022**, *271*, 108171. [CrossRef]
25. Parrinello, M.; Rahman, A. Polymorphic transitions in single crystals: A new molecular dynamics method. *J. Appl. Phys.* **1981**, *52*, 7182–7190. [CrossRef]
26. Martyna, G.J.; Tobias, D.J.; Klein, M.L. Constant pressure molecular dynamics algorithms. *J. Chem. Phys.* **1994**, *101*, 4177–4189. [CrossRef]
27. Tuckerman, M.E.; Alejandre, J.; López-Rendón, R.; Jochim, A.L.; Martyna, G.J. A Liouville-operator derived measure-preserving integrator for molecular dynamics simulations in the isothermal–isobaric ensemble. *J. Phys. A Math. Gen.* **2006**, *39*, 5629. [CrossRef]
28. Purja Pun, G.P.; Mishin, Y. Development of an interatomic potential for the Ni-Al system. *Philos. Mag.* **2009**, *89*, 3245–3267. [CrossRef]
29. Honeycutt, J.D.; Andersen, H.C. Molecular dynamics study of melting and freezing of small Lennard-Jones clusters. *J. Phys. Chem.* **1987**, *91*, 4950–4963. [CrossRef]
30. Faken, D.; Jónsson, H. Systematic analysis of local atomic structure combined with 3D computer graphics. *Comput. Mater. Sci.* **1994**, *2*, 279–286. [CrossRef]
31. Stukowski, A. Visualization and analysis of atomistic simulation data with OVITO—The Open Visualization Tool. *Model. Simul. Mater. Sci. Eng.* **2010**, *18*, 015012. [CrossRef]
32. Stukowski, A.; Bulatov, V.V.; Arsenlis, A. Automated identification and indexing of dislocations in crystal interfaces. *Model. Simul. Mater. Sci. Eng.* **2012**, *20*, 085007. [CrossRef]
33. Xu, S.; Deng, X. Nanoscale void nucleation and growth and crack tip stress evolution ahead of a growing crack in a single crystal. *Nanotechnology* **2008**, *19*, 115705. [CrossRef] [PubMed]
34. Zhou, X.W.; Johnson, R.A.; Wadley, H.N.G. Misfit-energy-increasing dislocations in vapor-deposited CoFe/NiFe multilayers. *Phys. Rev. B* **2004**, *69*, 144113. [CrossRef]
35. Černý, M.; Pokluda, J. Ideal tensile strength of cubic crystals under superimposed transverse biaxial stresses from first principles. *Phys. Rev. B* **2010**, *82*, 174106. [CrossRef]
36. Pei, L.; Lu, C.; Tieu, K.; Zhao, X.; Zhang, L.; Cheng, K. Ductile-to-brittle fracture transition in polycrystalline nickel under tensile hydrostatic stress. *Comput. Mater. Sci.* **2015**, *109*, 147–156. [CrossRef]
37. Yang, Z.; Zhang, G.; Luo, G.; Sun, X.; Zhao, J. Mechanical properties of gold twinned nanocubes under different triaxial tensile rates. *Phys. Lett. A* **2016**, *380*, 2674–2677. [CrossRef]
38. Yang, Y.; Wang, X.; Zhang, G.; Zhang, Y.; Yang, Z. Molecular dynamics simulations of single crystal copper nanocubes under triaxial tensile loading. *Comput. Mater. Sci.* **2017**, *138*, 377–383. [CrossRef]

crystals

MDPI

Article

Performance Analysis of 7075 Aluminum Alloy Strengthened by Cavitation Water Jet Peening at Different Scanning Speeds

Zijun Zhang [1], Yongfei Yang [1,*], Yong Gao [2], Gaowei Wang [1] and Weidong Shi [1,*]

[1] School of Mechanical Engineering, Nantong University, Nantong 226019, China
[2] Mechanical Maintenance Department, Nantong Cellulose Fibers Co., Ltd., Nantong 226007, China
* Correspondence: yyf2020@ntu.edu.cn (Y.Y.); wdshi@ujs.edu.cn (W.S.)

Abstract: Shot peening technologies can improve the performance of materials and extend the service life of parts. Cavitation water jet peening (CWJP) is a new shot peening technology that uses the shear effect of a high-speed jet in a submerged environment to produce a cluster of bubbles, thereby achieving a significant increase in impact performance. In this paper, CWJP was used to strengthen the surface of the 7075 aluminum alloy (Al7075), and the effects of CWJP with different scanning speeds on its microstructure and mechanical properties were investigated. The morphology evolution revealed that the plastic deformation of the sample surface became more serious and the surface roughness increased with the decrease in scanning speed. The distribution of residual stress and microhardness along the depth direction of the Al7075 was measured before and after the CWJP impact, and it was found that CWJP causes work hardening on the surface of the Al7075 and can convert the subsurface tensile stress of the Al7075 into compressive stress. The mechanism of grain refinement of the Al7075 during plastic deformation was systematically revealed. Therefore, CWJP can effectively improve the surface properties of the 7075 aluminum alloy.

Keywords: cavitation water jet peening; 7075 aluminum alloy; microhardness; residual stress; microstructure evolution

1. Introduction

Shot peening techniques have been widely used in industry to improve material properties, especially fatigue life. Traditional mechanical shot peening utilizes solid balls to impinge the surface of a part, producing a large plastic deformation and introducing residual compressive stresses on the surface of that part; thus, improving its resistance to crack initiation and propagation [1,2]. On the other hand, however, this technique is characterized by severe pollution and large surface roughness. The resultant larger roughness facilitates stress concentrations on the surface, causing a significant reduction in its fatigue resistance and uneven deformation of the treated specimens [3]. Therefore, traditional mechanical shot peening is increasingly unable to meet the processing requirements of modern industry.

In the past few decades, a series of new peening processes have been derived from mechanical peening in order to overcome these limitations, including laser shock peening (LSP), water jet peening (WJP) and cavitation water jet peening (CWJP). LSP can produce high dislocation density, significant grain refinement and high-peak residual compressive stress on the surface of parts; thus, effectively improving the mechanical properties of these parts [4–6]. However, it requires a long processing time and expensive equipment. WJP, which employs high-speed water droplets as the peening media, has achieved good results in introducing residual compressive stresses on metal surfaces. WJP has many advantages over conventional mechanical attrition treatments, such as small surface roughness and low industrial pollution [7–9]. However, the driving pressure requirement of the high-pressure pure water jet peening is more than 200 MPa, which is an order of magnitude higher than that required for CWJP, causing excessive energy consumption.

CWJP is an innovative jet technology that uses the shear effect of a high-speed jet in a submerged environment to produce cavitation bubbles in order to greatly improve the impact performance [10,11]. After an impingement by a cavitation water jet, the surface properties of materials can be improved to a certain extent, and a certain thickness of strengthening layer will be formed on the surface of the materials. The existence of a residual compressive stress field near the surface of a material can balance the tensile stress generated on the surface of that material during the stressing process, as well as inhibit the initiation of cracks on the surface and subsurface of the material, reduce the expansion speed of cracks and, finally, improve fatigue life [12,13]. Soyama et al. first investigated the cavitation water jet in terms of improving the surface performance of parts and components and found that the residual compressive stress on the surface of parts and components after a cavitation water jet treatment has been significantly improved. Since then, the application of shot peening to the surface treatment of materials has become more and more extensive [14–16]. Hutli et al. verified that the cavitation water jet is an effective method to improve the surface properties of materials, and the force generated by the cavitation can change the surface roughness of materials on the nanoscale and micron scale [17,18]. Tsuda et al. used CWJP to treat steel gear shafts with complex shapes and structures. By analyzing and comparing the residual stress field on the surface of the parts before and after peening, the results show that CWJP can effectively enhance the residual compressive stress value on the material surface [19]. Masataka et al. investigated the strain-hardening effect of CWJP. It was found that CWJP can improve the hardness of pure aluminum at the surface and subsurface layers and introduce pressurized residual stress to enhance the fatigue strength [20,21]. He et al. used CWJP to strengthen a TC4 titanium alloy. The experiments and mechanism studies were carried out by means of macroscopic and microscopic morphologies, metallographic structures, XRD analyses, microhardness and residual stress methods [22].

In this paper, CWJP was used to strengthen the surface of the 7075 aluminum alloy. The effects of CWJP at different scanning speeds on its morphology, roughness and mechanical properties were investigated. The distribution of the residual stress and microhardness along the depth direction were measured before and after the CWJP treatment. The microstructure evolution of the 7075 aluminum alloy (Al7075) during the CWJP was studied and discussed. The mechanism of grain refinement of the Al7075 during plastic deformation was systematically revealed. Therefore, CWJP can effectively improve the surface properties of the 7075 aluminum alloy.

2. Experimental Procedures

2.1. Materials and Methods

The CWJP device equipped with an Italian AR high-pressure piston pump (AR, Italian), which can provide a pressure for the jet of up to 50 MPa, a rated speed of 1450 r/min and a flow rate of 15 L/min, is shown in Figure 1a. The high-pressure pipeline is equipped with a pressure relief valve and a pressure gauge for safety. The piston pump pressure is 20 MPa, and an organ pipe nozzle with a flare is applied in this study, which is fixed on a three-axis sliding table.

The material used in this study was Al7075 (China), and the dimensions of the specimens were 100 mm × 100 mm × 3 mm. The chemical composition of the Al7075 is shown in Table 1. The specimens were heat-treated and polished before the CWJP. The annealing of these specimens was undertaken at a temperature of 510 °C for 3 h in atmosphere, after which these samples were placed at room temperature for a complete air-cooling. It should be noted that the specimens were fixed on a horizontal platform below the nozzle by a fixture, and the nozzle axis was perpendicular to the platform. During the CWJP process, the nozzle was moved along a defined path by controlling the slide, and the scanning speeds were 1, 0.5 and 0.1 mm/min. For convenience, the specimens treated with the CWJP at the scanning speeds of 0.1 mm/min, 0.5 mm/min and 1 mm/min were defined as 1.0_CWJP, 0.5_CWJP and 0.1_CWJP, respectively. Additionally, the original specimen without a CWJP treatment was marked as unCWJP. The detailed scanning position and peening area of the cavitation jet are shown in Figure 1b. After the CWJP treatment, the surface of the specimens showed craters and grain fragmentation layers, as shown in Figure 1c, indicating that CWJP would have effects on the surface roughness and grain refinement of the Al7075 specimens.

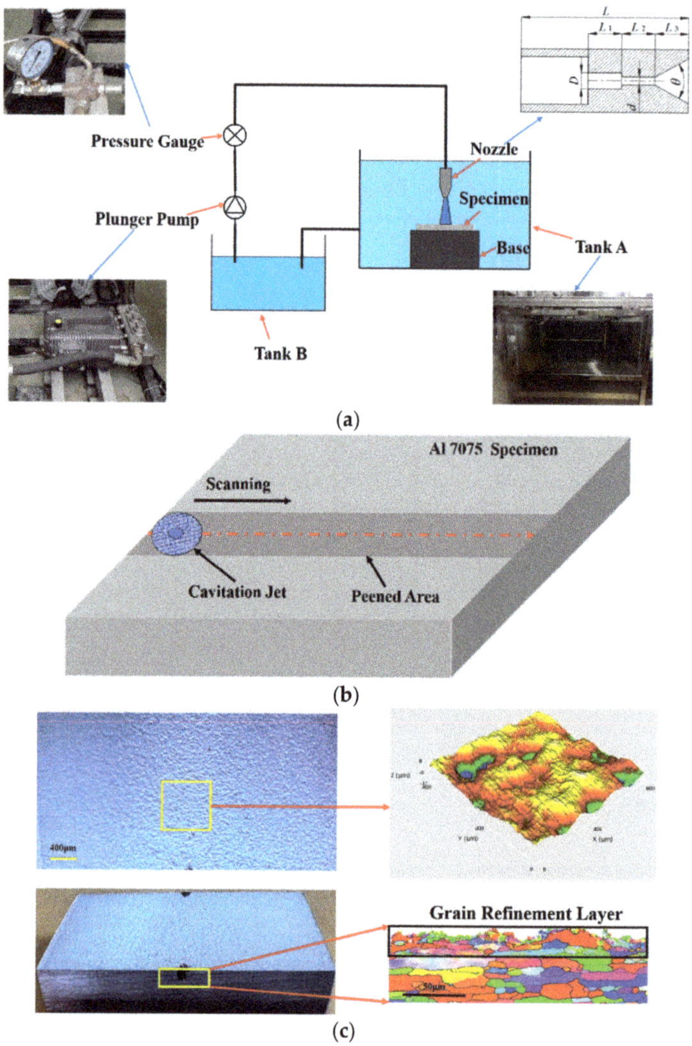

Figure 1. Schematic diagram of the CWJP process. (**a**) CWJP device; (**b**) scanning position and peening area; (**c**) surface of the specimen after the CWJP treatment.

Table 1. Chemical composition of Al7075.

Element	Zn	Mg	Cu	Mn	Ti	Fe	Zr	Si	Al
Wt.%	5.5	2.5	1.6	0.2	0.05	0.4	<0.1	<0.1	Bal.

2.2. Surface Topography and Roughness Measurement

After an CWJP treatment, the surfaces of the samples will undergo plastic deformation. In this experiment, the surface of the Al7075 after shot peening was observed by a Nanofocus Usurf confocal microscope (Nanofocus Usurf, Germany), and the geometric feature data were extracted using its own post-processing software. With the help of a scanning electron microscope (SEM), the surface topography of the samples before and after the treatment was observed.

2.3. Microhardness and Residual Stress Measurement

As a surface treatment technology, it is necessary to analyze the thickness of the surface hardening layer of the specimen. Measurements were performed on the microhardness trend of the specimen profile from the surface to the core, and the instrument used was a TWVS-1 digital display micro-Vickers hardness tester (TWVS-1, Shandong, China). The microhardness testing was carried out under a load of 0.981 N and a duration of 15 s to obtain the specimen hardness distribution profiles. Three points were measured at the same depth, and the average value was taken as the final hardness value under shot peening and non-shot peening conditions to acquire stable microhardness values.

An X-ray stress analyzer (Xstress 3000 G2R STRESSTECH OY, Jyväskylä, Finland) was used to measure the residual stress distribution on the surface of the impacted region. Five points were selected along the diameter of the impact crater, and each measurement point was measured once in each of the following three directions: 0°, 45° and 90°. The $\sin^2\psi$ method was applied and the corresponding test parameters were selected as follows: a collimated tube with a diameter of 1 mm; Cr as the target material; a Bragg angle of 156.4°; a crystal surface type of (211); a tube voltage of 30 kv; a tube current of 25 mA; an exposure time of 15 s; an elastic constant of 70,600.

2.4. Microstructural Characterization

After the CWJP treatment, the 0.5_CWJP specimens were selected for a metallographic study. The microstructure of the specimens after the CWJP was observed by transmission electron microscopy (TEM) (FEI Talos F200X, NY, USA) at a specific depth of 50 μm.

3. Results and Discussion

3.1. Surface Topography and Roughness Analysis

Figure 2 demonstrates the three-dimensional morphology of the surface of the Al7075 specimen at different scanning speeds, and Table 2 shows the corresponding surface roughness values. We measured each sample three times, as well as obtained its average value and standard deviation. Figure 2a shows the three-dimensional morphology of the surface of the specimen without impact passing, with a relatively flat surface and a roughness value of 0.46 μm. Figure 2b shows the three-dimensional surface morphology of the peened specimen at a scanning speed of 1 mm/min. It is found that the roughness value is about 1.77 μm. The high-pressure impact generated by the micro-jets and shock waves due to the collapse of the cavitation bubble on the surface of the specimen creates some plastic deformation craters on the surface of the specimen with small diameters and shallow depths, indicating the small amount of plastic deformation of the surface layer of the specimen. Thus, the peening intensity is not high nor effective in this working condition.

Figure 2c shows the three-dimensional morphology of the surface of the peened specimen at a scanning speed of 0.5 mm/min. With the increase in the impact duration, the corresponding roughness value rises remarkably to around 4.18 μm. The distribution density of the resulting craters increases to a certain extent, and both the diameter and the depth become larger compared to the ones in the case of 1.0_CWJP. When the scanning speed reaches 0.1 mm/min, the surface plastic deformation produced by the CWJP on the Al7075 specimen increases significantly. Therefore, as shown in Figure 2d, the roughness rises rapidly to 15.37 μm, implying the effectiveness of the CWJP on the Al7075.

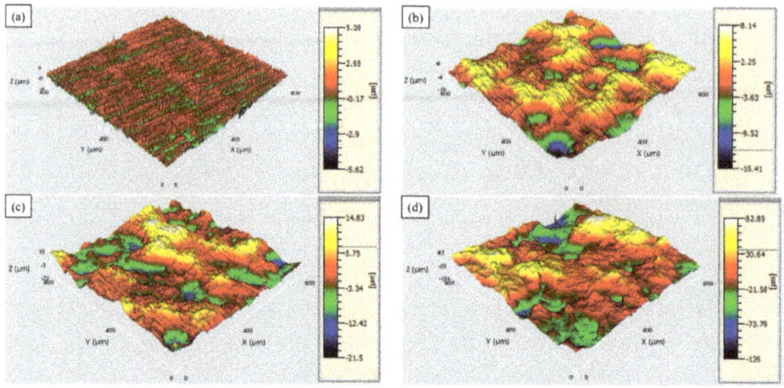

Figure 2. Three-dimensional morphology of the specimen surface at different scanning speeds. (**a**) unCWJP; (**b**) 1.0_CWJP; (**c**) 0.5_CWJP; (**d**) 0.1_CWJP.

Table 2. Surface roughness of specimens at different scanning speeds.

Specimen	unCWJP	1.0_CWJP	0.5_CWJP	0.1_CWJP
Sa/μm	0.46 ± 0.021	1.77 ± 0.122	4.18 ± 0.050	15.37 ± 0.342

The extensive plastic deformation on the surface of the specimen is formed by the impact owing to the collapse of a large number of cavitation bubbles; thus, the accumulation of treatment time is a key factor in the formation of effective plastic deformation. The SEM images in Figure 3 show the morphology of the sample surfaces at different scanning speeds. Figure 3a shows the SEM morphology of the specimen without an impact, and the surface of the specimen is relatively smooth except for the scratches. Figure 3b reveals the SEM morphology of the specimen at a scanning speed of 1 mm/min. Due to the fast scanning speed and the short action time, there are not enough cavitation bubbles to collapse; thus, the surface of the specimen cannot be deformed. Indeed, it can be observed that there are several pits on the surface of the sample, but the distribution density is not high.

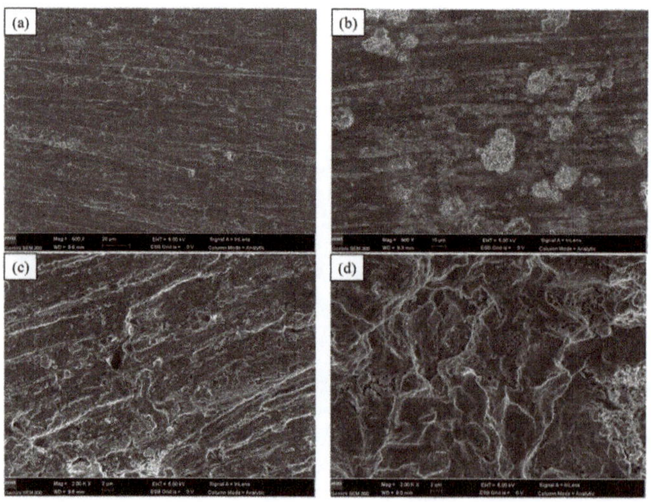

Figure 3. SEM morphology of specimen surfaces at different scanning speeds. (**a**) unCWJP, (**b**) 1.0_CWJP, (**c**) 0.5_CWJP, (**d**) 0.1_CWJP.

As shown in Figure 3c, the plastic deformation of the specimen increases as the scanning speed decreases to 0.5 mm/min. The impact craters show signs of edge buckling due to the pressure impact of the continuous collapse of the cavitation bubbles, and this extends to the periphery, showing a tendency for interconnection. From Figure 3d, it can be observed that with the further reduction in the cavitation jet peening scanning speed to 0.1 mm/min, the plastic deformation of the specimen surface further increases, as well as the area of depression caused by the jet impact expands over a large area and the number of craters caused by the collapse of the vacuole increases substantially; thus, the phenomenon of concatenation and overlap of craters appears.

3.2. Microhardness and Residual Stress Analysis

Figure 4 shows the magnitude of the microhardness values of the Al7075 at different depths before and after the CWJP treatment. The microhardness distribution of the unCWJP specimens in the depth direction is relatively stable, with an average value of 109.2 HV. After the CWJP treatment, grain refinement and dislocation strengthening occurred on the surface layer of the specimen, and the surface microhardness had the largest value. The CWJP scanning speed had a great influence on the surface microhardness of the specimen; the surface hardness value of the material tended to increase gradually as the scanning speed decreased, and the treatment time of the cavitation bubble became longer. When the CWJP scanning speed reached 1.0 mm/min, the highest microhardness of the specimen surface layer was 128.5 HV, and the microhardness of the specimen surface layer increased significantly compared with that of the unCWJP specimen. When the scanning speed of the CWJP was 0.5 mm/min, the maximum microhardness of the specimen surface layer was about 138.7 HV, and the increase in the specimen surface microhardness started to slow down. When the CWJP scanning speed was 0.1 mm/min, the maximum microhardness of the specimen surface layer was 144.0 HV, and the microhardness of the specimen surface layer increased more slowly. With the increase in the specimen depth, the microhardness gradually decreased until it stabilized at the same hardness value as the substrate, indicating that an effective hardened layer could be formed on the surface of the Al7075 after being reinforced by the CWJP; the CWJP scanning speed did not have much influence on the thickness of the hardened layer of the specimen, which was about 600 μm.

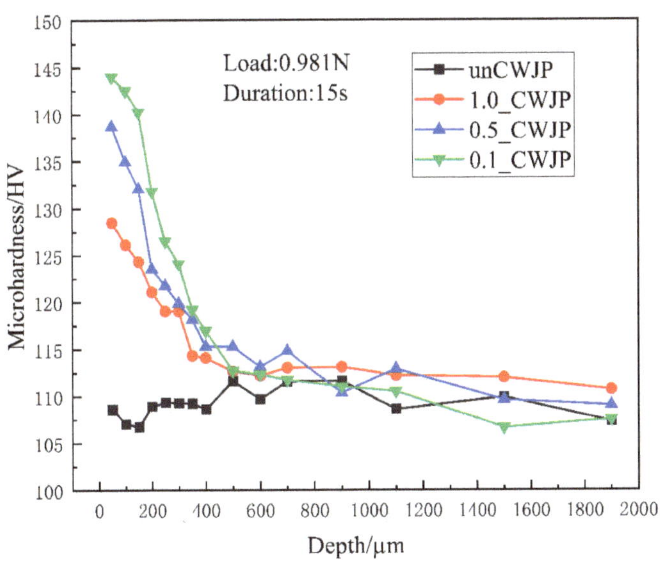

Figure 4. In-depth microhardness distribution curves.

Due to the presence of a large number of cavitation bubbles in the cavitation water jet, their collapse will turn the surroundings into a high-temperature and high-pressure environment, resulting in plastic deformation of the material surface, distortion of the internal lattice of the material, changes in the grain surface spacing and the appearance of dislocations, which are the internal factors of residual stress generation. Figure 5a shows the residual stresses in the surface layer of the Al7075 before and after the CWJP treatment. A high-magnitude tensile residual stress of 36.5 MPa was generated in the subsurface layer of the unCWJP specimen. During a CWJP, a severe plastic deformation of the material surface caused by the cavitation jet shock wave can lead to strain-hardening effects and implant compressive stresses in the material. When the CWJP scanning speed was 1.0 mm/min, the peak residual compressive stress in the surface layer of the specimen reached −136.9 MPa. When the CWJP scanning speed was 0.5 mm/min, the peak residual compressive stress of the specimen surface reached −176.9 MPa, and the maximum residual compressive stress of the specimen surface was elevated more gently. Additionally, when the CWJP scanning speed was 0.1 mm/min, the residual compressive stress in the surface layer of the specimen dropped rapidly to −88.8 MPa. The intrinsic reason is that the repeated impact of the cavitation bubble collapse at high pressure for a long time reaches the fatigue limit of the material, so that the reinforced layer originally formed on the surface of the material is stripped away by cavitation, which greatly reduces the surface properties of the material. Therefore, the CWJP time needs to be strictly controlled, and cavitation spalling should be strongly avoided after achieving the goal of improving the surface properties of the material. Considering the strengthening effect and strengthening efficiency, 0.5 mm/min was determined as the optimal scanning speed. Figure 5b shows the residual stress curve of the 0.5_CWJP specimen along the depth direction. The residual stress value from the surface of this specimen to a depth of 200 μm gradually became smaller along the depth; from 200 μm to 600 μm, the residual stress gradually became larger again. This indicates that the grain structure distribution is improved in a certain depth range after the CWJP surface treatment.

3.3. Microstructural Characterization Analysis

Figure 6 shows the TEM images of the microstructure at a 50-μm depth after the 0.5_CWJP. The results show that the main deformation mode is a dislocation slip. The plastic strain induces the production of high-density dislocation walls (Figure 6b), dislocation entanglement (Figure 6c) and dislocation cells (Figure 6d). From Figure 6a, it can be observed that only a few dislocation lines are randomly distributed in the substrate. When the material undergoes CWJP, a large number of dislocations appear, rearrange and annihilate to accommodate the plastic strain. As the strain increases, the stacked dislocation lines with the same sign pile up to produce dislocation walls and dislocation entanglement. The dislocation density is not uniform, and the dislocations within the grains are randomly distributed. Figure 6c depicts the complex entanglement between the dislocations. Figure 6d shows the full activation of the cross-slip, leading to high-density dislocation aggregation around the dislocation cell and sparse dislocation lines inside the dislocation cell.

Taheri-Nassaj et al. [23] pointed out that the clustered high-density dislocation structures interacted with each other and impeded their own motions, thereby contributing to a microhardness increase in this region. The strengthening mechanism was Taylor strengthening. Liu et al. [24] investigated the microstructural evolution of the FeCrNiCoMn alloy and believed that the average thickness of the mechanical twins (MTs) and their spacing played a significant role in the strengthening.

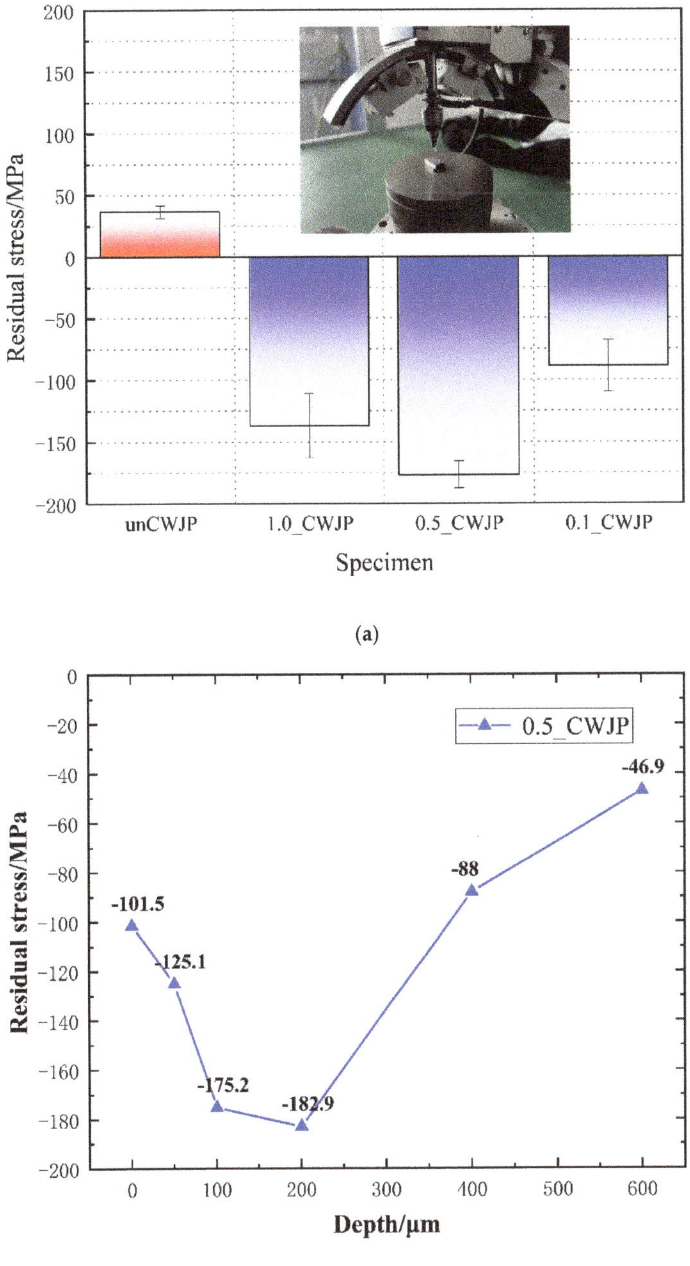

Figure 5. Residual stress. (**a**) Surface residual stress before and after CWJP. (**b**) Residual stress distribution curve of the 0.5_CWJP specimens in depth.

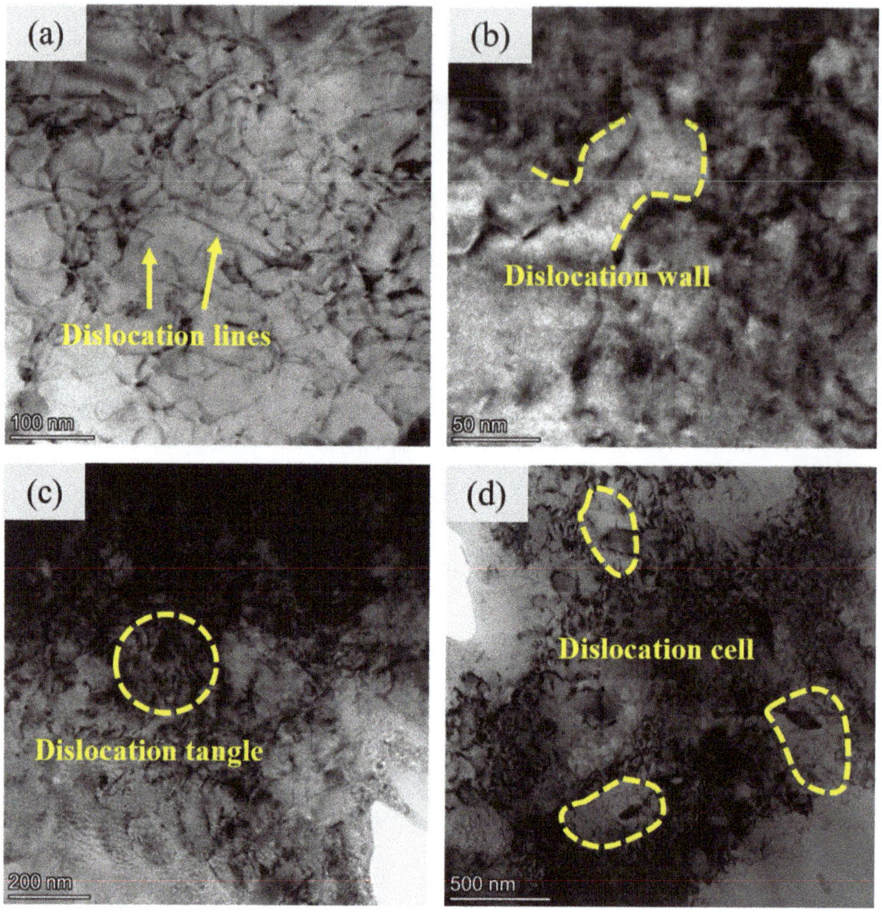

Figure 6. TEM images of the microstructure at a 50-depth after the 0.5_CWJP. (**a**) Dislocation lines; (**b**) dislocation walls; (**c**) dislocation tangles; (**d**) dislocation cells.

In the paper, on the other hand, mechanical twins cannot be observed in Al alloys even though they are under high strain rates due to the cavitation impact. The high stacking fault energy and dislocation slip of the Al7075 lead to the occurrence of dynamic recrystallization processes, and in the near-surface region, severe plastic strain refines the coarse grains to ultra-fined grains, which also contributes significantly to the corresponding maximum microhardness. Stepanov et al. [25] proposed that substructure hardening should be taken into account due to the dislocation density. The overall hardness is associated with the microhardness of the substrate, the substructure hardening and the Hall–Petch hardening, indicating that dislocation and grain size are the critical factors.

The microhardness of the cavitation jet-strengthened Al7075 containing dislocation strengthening and grain refinement strengthening is described as follows:

$$HV = HV_0 + M\alpha Gb\sqrt{\rho} + \frac{K_y}{\sqrt{d}} \tag{1}$$

From the equation, it can be observed that the higher the dislocation density, the greater the dislocation strengthening effect. From the TEM diagram, it can be observed that the material is deformed after the impact, and a large number of dislocations are entangled to form a high dislocation density. The dense high-density dislocation structures interact with each other and hinder their own motion; thus, promoting the increase in microhardness. The hardness of the material increases as the grain size decreases; therefore, grain refinement also plays an important role in the cavitation jet strengthening of the 7075 aluminum alloy.

3.4. Grain Refinement Mechanisms Induced by CWJP

When the impact source acts on the specimen, a shock wave is formed on the surface of the material. The resultant propagation of the shock wave inside the material induces the intracrystalline dislocations to move. The grain refinement process of the Al7075 during the CWJP can be divided into five steps, which are described as follows, and the schematic diagram is illustrated in Figure 7.

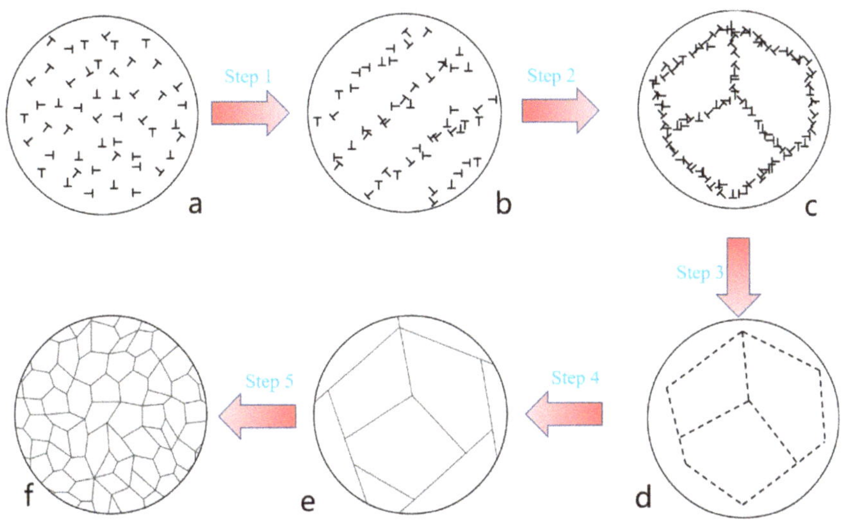

Figure 7. Schematic diagram of grain refinement processes of the Al7075 induced by CWJP. (**a**) Dislocation; (**b**) Dislocation lines; (**c**) Dislocation walls and dislocation tangles; (**d**) Subgrains; (**e**) DRX nucleation; (**f**) Refined grains.

Step 1: The dislocation behavior leads to the formation of dislocation lines within the original coarse crystal.

Step 2: The accumulation of dislocation lines leads to the formation of dislocation walls and dislocation tangles.

Step 3: As the strain increases further, the individual dislocation cells, initially separated by the dislocation entanglement and dislocation walls, will gradually subdivide into the original subgrains.

Step 4: When the deformation strain reaches a certain level, that is, when the dislocation density of the dislocation walls and dislocation tangles reaches a certain value, the annihilation of the rearrangement of these dislocations forms low angular grain boundaries and subdivides the original coarse grains into different subgrains.

Step 5: The high-level stacking fault energy and the dislocation slip of the aluminum alloy lead to a continuous dynamic recrystallization process, which leads to a further increase in grain boundary misorientation. Finally, the grain boundary properties are changed gradually until the formation of high-angle grain boundaries.

4. Conclusions

In this paper, CWJP was used to strengthen the surface of the 7075 aluminum alloy, and the effects of the CWJP at different scanning speeds on its microstructure and mechanical properties were investigated. The surface morphology and roughness of the Al7075 under the CWJP process at different scanning speeds were observed. The distribution of the microhardness and residual stress along the thickness direction before and after the CWJP impact were measured. The microstructure evolution of the Al7075 during the CWJP was studied and discussed. The mechanism of grain refinement of the Al7075 during plastic deformation was systematically revealed. The specific conclusions are summarized below.

(1) CWJP causes plastic deformation on the surface of the Al7075 and increases the surface roughness. As the scanning speed decreases and the treatment time of the shot peening becomes longer, the plastic deformation on the surface of the specimen becomes more serious, as well as the phenomenon of concatenation and overlapping of pits appears and the surface roughness increases.
(2) CWJP will produce working hardening on the surface of the Al7075. The thickness of the hardened layer is about 600 μm, and the maximum hardness of the specimen surface gradually increases as the scanning speed decreases. The CWJP treatment can convert the surface tensile stress of the Al7075 into compressive stress. However, when the shot peening time is too long, the originally formed strengthening layer on the material surface will be cavitationally stripped away; thus, the CWJP time needs to be strictly controlled.
(3) The main deformation mode of the CWJP impact on the Al7075 is the dislocation slip, which will lead to grain refinement of the Al7075 with high stacking fault energy. The specific process is as follows: Firstly, dislocation lines are formed in the original coarse crystal; the accumulation of dislocation lines leads to the formation of dislocation walls and dislocation tangles; the dislocation walls and dislocation tangles subdivide the coarse crystal into subgrains; finally, under the action of an external load, the subgrains are dynamically recrystallized into high-angle grain boundaries and new grains are formed.

Author Contributions: Data curation, Y.Y. and Y.G.; formal analysis, Z.Z., Y.Y. and G.W.; funding acquisition, Y.G. and W.S.; resources, W.S.; software, Z.Z. and G.W.; supervision, W.S.; writing—original draft, G.W.; writing—review and editing, Y.Y. and G.W. All authors have read and agreed to the published version of the manuscript.

Funding: This research was funded by the National Key Research and Development Project of China (No. 2019YFB 2005300), National High-Tech Ship Scientific Research Project of China (No. MIIT [2019] 360), National Natural Science Foundation of China (No. 51979138), National Natural Science Foundation of China (No. 273746), National Natural Science Foundation of China (No. 51979240), Jiangsu Natural Science Research Project (No. 19KJB470029), China Postdoctoral Science Foundation (No. 273746), and Natural Science Foundation of Jiangsu Province (No. BK20220609), Project funded by China Postdoctoral Science Foundation (No. 2022TQ0127).

Institutional Review Board Statement: Not applicable.

Informed Consent Statement: Not applicable.

Data Availability Statement: Data on the analysis and reporting results during the study can be obtained by contacting the authors.

Conflicts of Interest: The authors declare no conflict of interest.

References

1. Haghighi, O.; Amini, K.; Gharavi, F. Effect of shot peening operation on the microstructure and wear behavior of AZ31 magnesium alloy. *Prot. Met. Phys. Chem. Surf.* **2020**, *56*, 164–168. [CrossRef]
2. Azar, V.; Hashemi, B.; Yazdi, M.R. The effect of shot peening on fatigue and corrosion behavior of 316L stainless steel in Ringer's solution. *Surf. Coat. Technol.* **2010**, *204*, 3546–3551. [CrossRef]
3. Fargas, G.; Roa, J.J.; Mateo, A. Effect of shot peening on metastable austenitic stainless steels. *Mater. Sci. Eng. A* **2015**, *641*, 290–296. [CrossRef]
4. Jing, Y.; Fang, X.; Xi, N.; Feng, X.; Huang, K. Investigation of microstructure and mechanical properties evolution in 7050 aluminum alloy and 316L stainless steel treated by laser shock peening. *Mater. Charact.* **2021**, *182*, 111571. [CrossRef]
5. Kaufman, J.; Spirit, Z.; Vasudevan, V.K.; Steiner, M.A.; Mannava, S.R.; Brajer, J.; Pina, L.; Mocek, T. Effect of laser shock peening parameters on residual stresses and corrosion fatigue of AA5083. *Metals* **2021**, *11*, 1635. [CrossRef]
6. Wang, J.T.; Xie, L.; Luo, K.Y.; Tan, W.S.; Cheng, L.; Chen, J.F.; Lu, Y.L.; Li, X.P.; Ge, M.Z. Improving creep properties of 7075 aluminum alloy by laser shock peening. *Surf. Coat. Technol.* **2018**, *349*, 725–735. [CrossRef]
7. Cao, Y.-P.; Cheng, S.-M.; Shi, W.-D.; Yang, Y.-F.; Wang, G.-W. Experimental study on surface erosion of grade a marine steel by ultrahigh-pressure water jet. *Water* **2022**, *14*, 1953. [CrossRef]
8. Azhari, A.; Schindler, C.; Nkoumbou, J.; Kerscher, E. Surface erosion of carbon steel 1045 during waterjet peening. *J. Mater. Eng. Perform.* **2014**, *23*, 1870–1880. [CrossRef]
9. Barriuso, S.; Lieblich, M.; Multigner, M.; Etxeberria, I.; Alberdi, A.; Gonzalez-Carrasco, J.L. Roughening of metallic biomaterials by abrasiveless waterjet peening: Characterization and viability. *Wear* **2011**, *270*, 634–639. [CrossRef]
10. Soyama, H. Cavitating jet: A review. *Appl. Sci.* **2020**, *10*, 7280. [CrossRef]
11. Soyama, H. Cavitation peening: A review. *Metals* **2020**, *10*, 270. [CrossRef]
12. Yang, Y.; Li, W.; Shi, W.; Wang, C.; Zhang, W. Experimental study on submerged high-pressure jet and parameter optimization for cavitation peening. *Mechanika* **2020**, *26*, 346–353. [CrossRef]
13. Yang, Y.; Li, W.; Shi, W.; Zhou, L.; Zhang, W. Experimental study on the unsteady characteristics and the impact performance of a high-pressure submerged cavitation jet. *Shock Vib.* **2020**, *2020*, 1701843. [CrossRef]
14. Soyama, H.; Korsunsky, A.M. A critical comparative review of cavitation peening and other surface peening methods. *J. Mater. Process. Technol.* **2022**, *305*, 117586. [CrossRef]
15. Soyama, H.; Takeo, F. Comparison between cavitation peening and shot peening for extending the fatigue life of a duralumin plate with a hole. *J. Mater. Process. Technol.* **2016**, *227*, 80–87. [CrossRef]
16. Soyama, H.; Simoncini, M.; Cabibbo, M. Effect of cavitation peening on fatigue properties in friction stir welded aluminum alloy AA5754. *Metals* **2020**, *11*, 59. [CrossRef]
17. Hutli, E.A.F.; Nedeljkovic, M.S.; Radovic, N.A. Nano- and micro-scale surface modification of fcc metal using high submerged cavitating water jet. *Plasmonics* **2013**, *8*, 843–849. [CrossRef]
18. Hutli, E.; Bonyar, A.; Oszetzky, D.; Nedeljkovic, M.S. Plastic deformation and modification of surface characteristics in nano- and micro-levels and enhancement of electric field of FCC materials using cavitation phenomenon. *Mech. Mater.* **2016**, *92*, 289–298. [CrossRef]
19. Tsuda, H.; Ju, D.Y.; Uchiyama, T.; Sunayama, Y.; Oba, R. Residual stress in a quenched gear shaft treated by water cavitation peening. *Mater. Sci. Forum* **2005**, *490*, 364–369. [CrossRef]
20. Ijiri, M.; Shimonishi, D.; Nakagawa, D.; Yoshimura, T. New water jet cavitation technology to increase number and size of cavitation bubbles and its effect on pure Al surface. *Int. J. Lightweight Mater. Manuf.* **2018**, *1*, 12–20. [CrossRef]
21. Ijiri, M.; Shimonishi, D.; Tani, S.; Okada, N.; Yamamoto, M.; Nakagawa, D.; Tanaka, K.; Yoshimura, T. Multifunction cavitation technology to improve the surface function of Al–Cu alloy. *Int. J. Lightweight Mater. Manuf.* **2019**, *2*, 50–56. [CrossRef]
22. He, P.; Li, F.; Guo, J.; Chen, S.; Guo, Y.; Wang, Y.; Tan, Z. Research on the water cavitation peening process and mechanism of TC4 titanium alloy. *Int. J. Adv. Manuf. Technol.* **2021**, *112*, 1259–1269. [CrossRef]
23. Taheri-Nassaj, N.; Zbib, H.M. On dislocation pileups and stress-gradient dependent plastic flow. *Int. J. Plast.* **2015**, *74*, 1–16. [CrossRef]
24. Liu, W.H.; Wu, Y.; He, J.Y.; Nieh, T.G.; Lu, Z.P. Grain growth and the Hall–Petch relationship in a high-entropy FeCrNiCoMn alloy. *Scr. Mater.* **2013**, *68*, 526–529. [CrossRef]
25. Stepanov, N.; Tikhonovsky, M.; Yurchenko, N.; Zyabkin, D.; Kilmova, M.; Zherebtsov, S.; Efimov, A.; Salishchev, G. Effect of cryo-deformation on structure and properties of CoCrFeNiMn high-entropy alloy. *Intermetallics* **2015**, *59*, 8–17. [CrossRef]

crystals

MDPI

Article

Numerical Simulation and Process Study on Laser Shock Peening of 1Cr18Ni9Ti Material

Xiangyu Ding [1,2], Sijie Ma [2], Junlong Zhang [2], Zonghong Jiang [2], Hongliang Li [2], Shengchao Wang [2], Cheng Wang [1] and Jida Zhong [2,*]

1 Fundamentals Department, Air Force Engineering University, Xi'an 710051, China
2 School of Aircraft Engineering, Nanchang Hangkong University, Nanchang 330063, China; zjl082100@163.com (J.Z.)
* Correspondence: z1526293936@163.com

Abstract: This article mainly studies the improvement of the properties of the 1Cr18Ni9Ti material after laser shock peening. The 1Cr18Ni9Ti material is the main material used to make aviation ducts, and improving the fatigue life of aviation ducts can significantly improve the safety performance of aviation engines. The article combines simulation and experiment to study the improvement effect of laser shock peening on the material's properties. The main results are as follows: The fatigue test showed that, under the same stress load, laser shock peening can greatly extend the fatigue life of the specimen, with the 3J process having the best effect. EBSD analysis showed that the 3J process has the best grain refinement effect. The X-ray diffraction method proved that the measurement results of residual compressive stress under the 3J process are optimal. Overall, it is shown that the properties of the 1Cr18Ni9Ti material can be greatly improved under the 3J process.

Keywords: laser shock peening; process research; welding residual stress; aero duct; numerical simulation

1. Introduction

Aero engines play an important role in the military industry, and there are many welding links in the manufacturing process. Once parts are damaged, it is difficult to achieve the original performance of the material by maintenance and repair welding [1]. Aviation conduits, known as the blood vessels of aircraft, are an important part of the aircraft's control system, with important roles such as transferring pressure, delivering fuel, and protecting electrical cables. While aviation conduits rely mainly on welding for connection, the strength of welded conduit joints is low, coupled with the poor working environment of aviation engines, so aviation conduits are prone to rupture during operation, which has great safety hazards for the operation of aviation engines.

Welding is the main form of joining modern steel. Its advantages are that the member section is not weakened, the assembly plate can be omitted, the structure is simple, the steel is saved, the manufacturing and processing are convenient, and the sealing performance is good. However, the local temperature at the joint of the weldment is higher; the forming quality of the heat-affected zone material is poor; the cooling is fast; the heat-affected zone shrinks unevenly, which easily results in residual stress in the welding seam, residual deformation of the welding seam, and even cracks, resulting in brittle fracture. Laser shock peening (LSP) has been developed rapidly in the field of welding due to its advantages such as no introduction of foreign medium, flexible laser transmission, good surface condition after strengthening and a deeper residual compressive stress layer. Logesh [2] and Leo [3] conducted research on weld strengthening and found that laser shock peening can effectively enhance the weld strength values. Chattopadhyay [4], Feng [5], and Shi [6] found that laser shock peening enhances the fatigue limit of the welding of different materials.

The principle of laser shock peening is illustrated in Figure 1 below. An absorbent protective layer is covered on the surface of the material, which acts on the surface of the material using high-power-density (greater than 1 GW/cm^2) and short pulses (nanosecond order) of laser light. The absorbent protective layer is capable of rapidly absorbing the laser energy, producing an explosive vaporization reaction, and forming a plasma at a high temperature (about 10^4 K) and a high pressure (greater than 1 GPa), which ejects into the surface of the metal under the confinement layer, thus inducing intense high-pressure shock stress waves. When a high-pressure impact stress wave propagates into a material's interior, a plastic deformation layer forms on the material if its peak pressure is between the dynamic yield strength and the tensile strength of the material. The high-density dislocations and surface residual compressive stresses in this plastic layer are combined to improve the mechanical properties of the material [7]. Compared to the typical shot-peening procedure, the residual compressive stress layer that is generated has a significant depth and has little influence on the surface roughness [8–10]. The greater the roughness, the more significant the decrease in the fatigue life is [11]. Many experts in this field have also confirmed that laser shock peening improves the fatigue limit of various different materials [12–15].

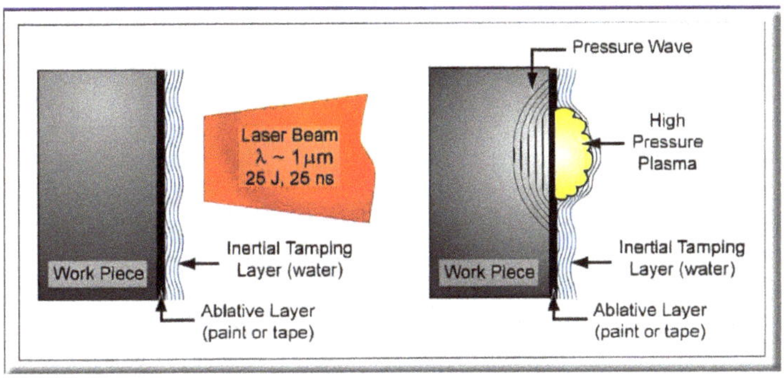

Figure 1. Principles of laser shock peening.

2. Materials and Methods

2.1. Simulation Analysis of Impact Strengthening of Fatigue Test Article

Due to the extremely short nanosecond action time of the laser shock, the action process of the shock wave cannot be observed by conventional experimental means. The complete process of shock wave action can be simulated by the finite element method, so finite element simulation is an effective means to study the laser shock processing parameters. Modelling in this article refers to the standard GBT4337-2015 for fatigue testing of metallic materials. The selected dimensions are shown in Figure 2, and the established model is shown in Figure 3 (in mm).

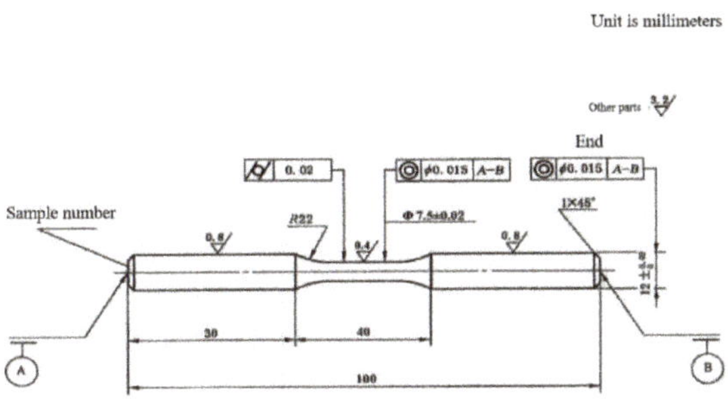

Figure 2. Fatigue test standard for metal materials GBT4337-2015.

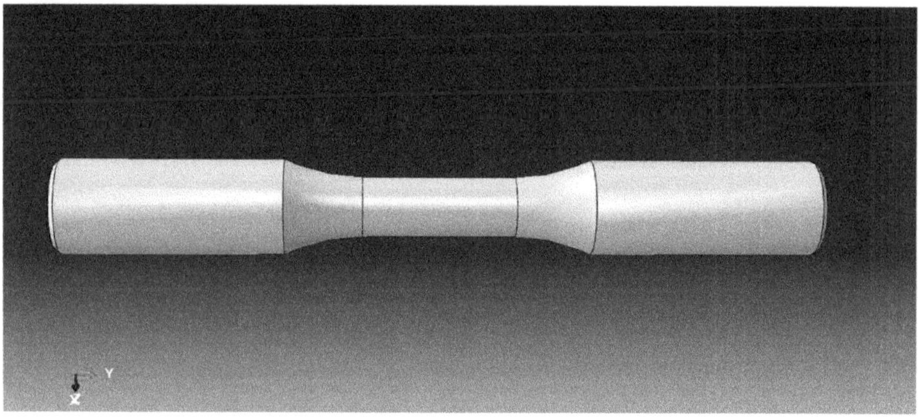

Figure 3. ABAQUS simulation model.

A large number of constituent models have been proposed to describe the dynamic response processes of materials, such as the power-law (PL) [16] model, Steinberg–Guinan (SG) [17] model, Zerilli–Armstrong (ZA) [18] model, Johnson–Cook (JC) [19] model, etc. In the Johnson–Cook model, the strain, strain rate, and temperature are taken into account, so the JC model was chosen for the simulation, which is also the parameter model that is being used in the field of laser shock peening at home and abroad.

$$\sigma = (A + B\varepsilon^n)\left[1 + C\ln(\dot{\varepsilon}^*)\right]\left[1 - (T^*)^m\right]$$

where A is the yield strength of the material; B is the working hardening modulus; N is the coefficient of hardening; C reflects the strain rate hardening effect of the material; ε is the plastic strain; $\dot{\varepsilon}^*$ is the dimensionless plastic strain rate; $T^* = (T - T_0)/(T_m - T_0)$ is the dimensionless temperature; T_0 is room temperature; T_m is the melting point of the material; and m reflects the temperature-weakening effect of the material. The material JC model parameters are shown in Table 1 [20].

Table 1. 1Cr18Ni9Ti material's property parameters and Johnson–Cook model parameters.

Density ρ ($kg{\cdot}m^{-3}$)	Modulus of Elasticity E (MPa)	Poisson Ratio μ	Melting Point of Material T (°C)	A (MPa)	B (MPa)	C	n	m
7930	206,000	0.27	1446	224	857	0.01	0.208	1.7855

2.1.1. Grid and Boundary Conditions Setting

The grid uses a hexahedral grid with a size of 0.2 mm × 0.2 mm, a total number of grids of 305,694. To better relate simulations to experiments, the boundary settings are consistent with experimental processing. Figure 4 shows the boundary settings. Arrows indicate the direction of laser impact; the red area of the specimen represents the laser strengthened area; the orange part represents the clamping area of the specimen, which is represented by multiple points for easy identification and viewing. Figure 5 shows the impact path of the S type.

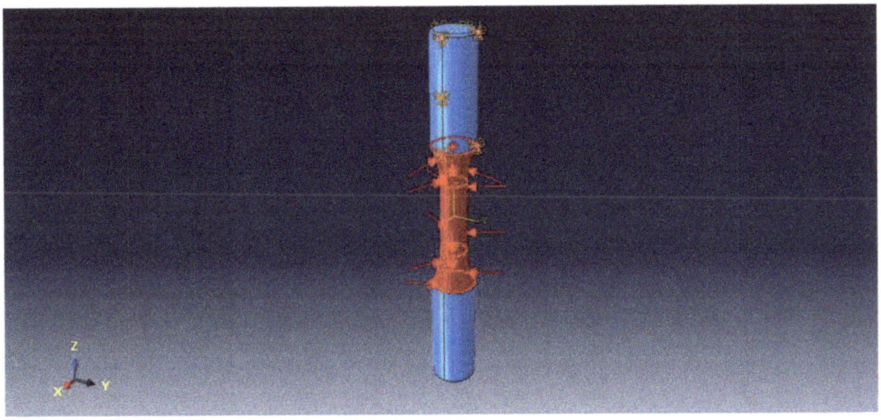

Figure 4. Loads and boundary conditions schematics.

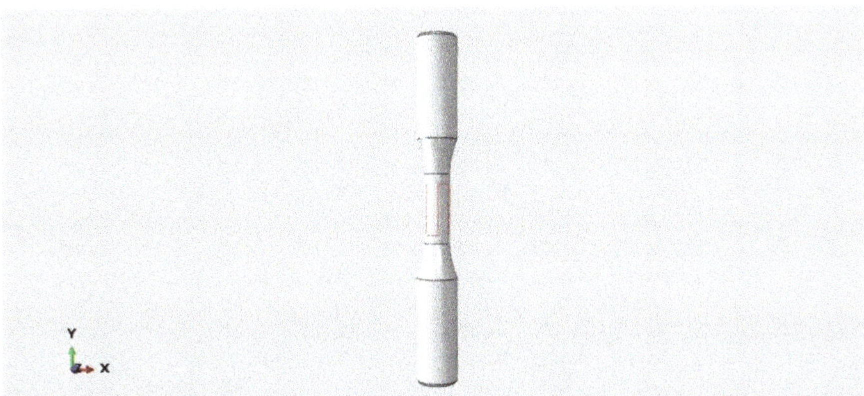

Figure 5. Processing path diagram.

2.1.2. Laser Parameter Selection

Normal stress at the yield of the material under one-dimensional strain impact compression σ_x is the Hugoniot elastic limit, denoted as σ_{HEL} [21].

$$\sigma_x = \sigma_{HEL} = K\varepsilon_{HEL} + \frac{2}{3}Y_0 = \left(K + \frac{4}{3}G\right)\varepsilon_{HEL} \tag{1}$$

where Y_0 is the Tresca yield stress of the material, K is the bulk modulus, and G is the shear modulus. The modulus of elasticity and Poisson's ratio of a material are as follows:

$$K = \lambda + \frac{2}{3}\mu = \frac{E}{3(1-2\nu)} \tag{2}$$

$$G = \mu = \frac{E}{2(1+\nu)} \tag{3}$$

The following formula is then obtained:

$$\sigma_{HEL} = \left(\frac{K}{2G} + \frac{2}{3}\right)Y_0 \tag{4}$$

According to the shock wave peak pressure P, the material Hugoniot elastic limit σ_{HEL} gives the amount of plastic deformation on the surface of metal parts strengthened by laser shock [22]:

$$\varepsilon_P = \frac{-2\sigma_{HEL}}{3\lambda + 2\mu}\left(\frac{P}{\sigma_{HEL}} - 1\right) \tag{5}$$

As shown in Figure 6, when $P < \sigma_{HEL}$, there is only elastic deformation; when $\sigma_{HEL} < P < 2\sigma_{HEL}$, plastic deformation occurs, accompanied by elastic recovery, which increases linearly with the impact peak pressure P; when $2\sigma_{HEL} < P < 2.5\sigma_{HEL}$, plastic deformation is saturated, reaching and maintaining a maximum of $-2\sigma_{HEL}/(3\lambda + 2\mu)$; when $P > 2.5\sigma_{HEL}$, the residual compressive stress on the metal surface is somewhat reduced due to the surface unloading wave. Therefore, the theoretical optimal peak pressure for laser shock peening should be taken as $P = 2\text{~}2.5\sigma_{HEL}$.

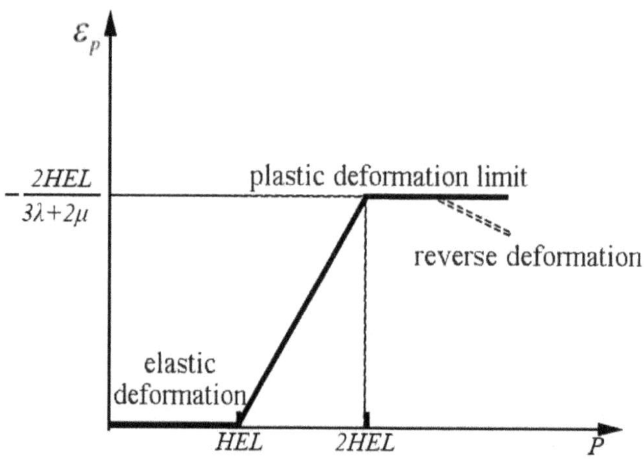

Figure 6. A diagram of metal elastoplastic deformation as a function of impact peak pressure.

2.1.3. Analysis of Simulation Results

Simulations set up multipoint continuous impacts, similar to the actual machining mode, with each impact point spaced 5000 ns apart for dynamic stabilization time to achieve stabilization at each impact point, while precluding mutual interference between each impact point. A 50% overlap rate was achieved between impact points to ensure the strengthening effect, and the strengthening path is shown in Figure 7. The impact points were 5 in each column, with a longitudinal overlap in the circumference and a spot diameter of 2 mm, and the peak pressure was the control variable, uniformly strengthening for one week.

According to the previous calculation, the theoretically most-suitable peak impact pressure is controlled between 2539 MPa and 3173 MPa. If a smaller pressure is selected to convert it into laser parameters, this parameter exceeds the lower limit of the experimental equipment parameters. If the difference in peak pressure selection is too small, it is difficult to compare the enhancement effect. Therefore, peak pressures of 3000 MPa, 4000 MPa, and 5000 MPa were selected for analysis and calculation in this section to verify the theoretical calculation parameters. The simulation results from left to right peak pressures of 3000 MPa, 4000 MPa, and 5000 MPa, respectively, are as follows:

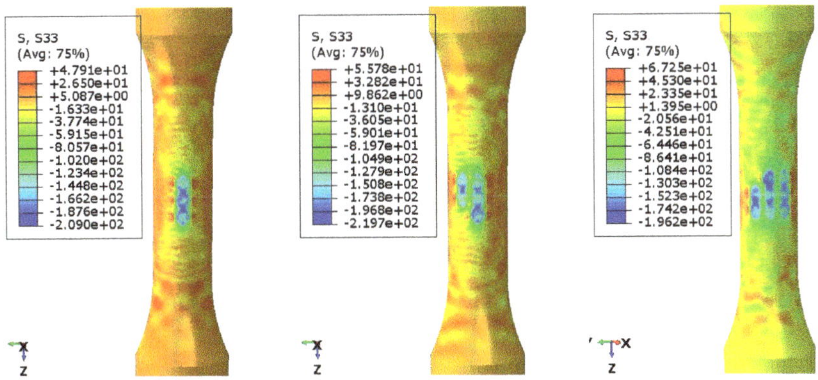

Figure 7. A diagram of the impact-strengthening process.

From Figure 8, it can be seen that in the S33 direction, the surface residual stress generated under the 3000 MPa parameter is the smallest, about 278 MPa compressive stress, while the surface residual stress generated under 4000 MPa is 393 MPa, and the surface residual stress generated under 5000 MPa is 341 MPa. However, the surface stress distribution of the material is the most uniform under the 3000 MPa parameter.

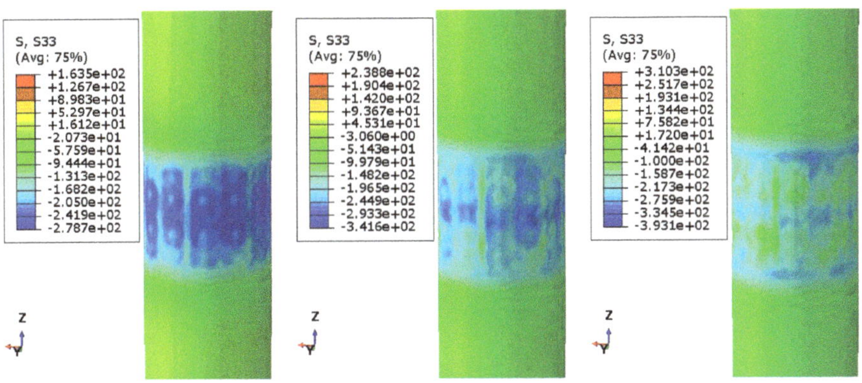

Figure 8. Cloud diagram of surface residual stress with different parameters.

The material plastic deformation simulation results are shown in Figure 9. Plastic deformation can cause damage to the material's surface roughness, which can be reduced to damage the material's properties, especially for some precision parts such as turbine blades, which have extremely high requirements for the material's surface roughness. Cloud charts show that the local shaping deformation of the material is minimal and the surface roughness is highest when the peak pressure is 3000 MPa.

The rebound of shock waves while residual compressive stresses form on the surface of a part strengthened by laser shock can cause tensile stresses in the interior of the part, and the magnitude and depth of residual tensile stresses present on the subsurface under different processes vary, as shown in Figure 10. The maximum tensile stresses all occur near the center of the circle, 163 MPa at 3000 MPa, 238 MPa at 4000 MPa, and 309 MPa at 5000 MPa. Larger internal tensile stresses can have a large negative effect on the life of the material, most likely exacerbating the propagation of internal cracks, so it can be seen that unsuitable machining parameters can have counterproductive effects.

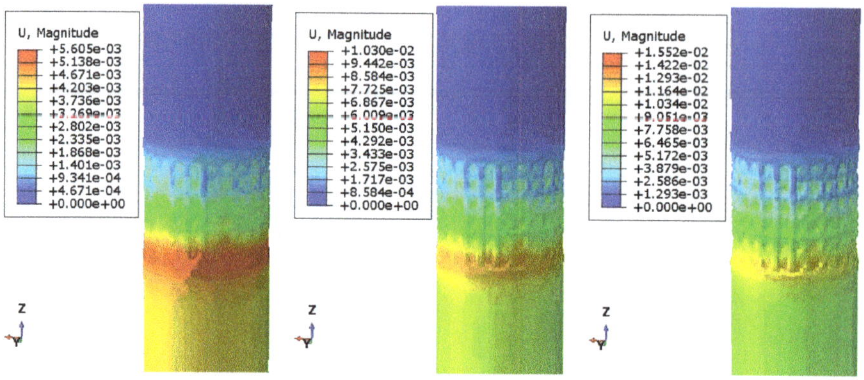

Figure 9. Enlarged cloud map of surface deformation at different parameters.

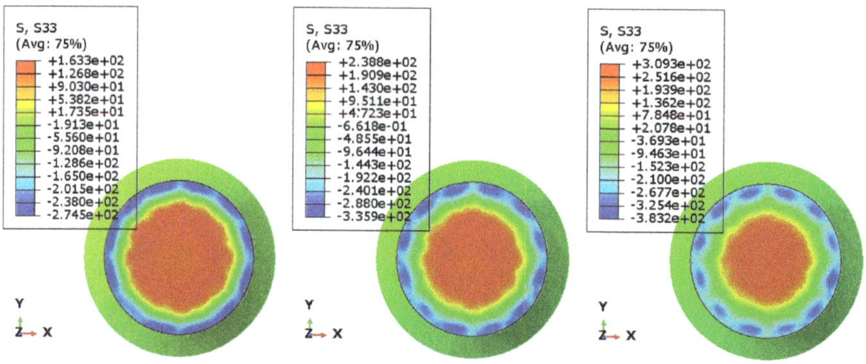

Figure 10. Stress cloud in internal tensile direction under different parameters.

2.1.4. Comprehensive Results and Analysis

The simulation results are summarized in Table 2:

Table 2. Summary of simulation results.

Peak Pressure	Maximum Residual Compressive Stress on the Surface	Thickness of Residual Compressive Stress Layer	Internal Maximum Tensile Stress	Mean Deformation
3000 MPa	−274.5 MPa	0.83 mm	163.3 MPa	3.26 μm
4000 MPa	−335.9 MPa	0.96 mm	238.8 MPa	6.00 μm
5000 MPa	−383.2 MPa	1.04 mm	309.3 MPa	9.05 μm

The conversion of the simulation parameters to the actual machining parameters is based on the macro equations also proposed by Fabbro:

$$P(\mathrm{GPa}) = 0.01 \cdot \sqrt{\frac{\alpha}{\alpha+3}} \cdot \sqrt{Z(\mathrm{g{\cdot}cm^{-2}{\cdot}s^{-1}}) \cdot \sqrt{A \cdot I(\mathrm{GW{\cdot}cm^{-2}})}} \tag{6}$$

In the formula above, α is the process efficiency, typically between 0.25 and 0.4, A is the average absorption, between 0.75 and 0.9, Z is the target and confinement layer acoustic impedance, and I is the laser power density. Their specific values are shown in Table 3:

Table 3. Conversion table of impact peak pressure and processing parameters.

P	α	Z1 (Target Material)	Z2 (Constrained Layer)	Z	d	τ	E	I
3093 MPa 4077 MPa 5030 MPa	0.2	4.34×10^6 g·cm^{-2}·s^{-1}	1.65×10^5 g·cm^{-2}·s^{-1}	3.18×10^5 g·cm^{-2}·s^{-1}	2 mm	20 ns	3J 5J 8J	4.77 Gw/cm^2 7.32 Gw/cm^2 11.1 Gw/cm^2

Comparing the results of the three peak pressure impacts, an average deformation of 3.26 μm is generated at 3000 MPa, while 6.00 μm and 9.05 μm are generated at 4000 MPa and 5000 MPa, respectively, which shows that the deformation of the surface is getting bigger and bigger with the increase of the peak pressure. Here, 3000 MPa has the most homogeneous distribution of the stresses, and the internal tensile stresses are small, but the relative layer of the residual compressive stresses is thinner and the surface residual stress the smallest at 3000 MPa, which is about 278 MPa and 393 MPa, respectively. The residual stress on the surface at 3000 MPa is the smallest, about 278 MPa, which is less than the corresponding 341 MPa and 393 MPa at 4000 MPa and 5000 MPa. From the simulation results, it can be seen that the maximum residual compressive stress on the surface and the depth of the residual compressive stress layer under the peak pressure of 3000 MPa are slightly smaller than the other two parameters, but the residual tensile stress and deformation caused by it are much smaller than the other two parameters. The above results are all factors that affect the fatigue life of the specimen. In summary, the simulation results showed that the parameter of 3000 MPa is the best for improving the fatigue life of the specimen. Finally, the actual processing parameters were calculated from the peak impact pressure according to Fabbro's semi-empirical formula (Formula (6)), and the converted parameters were 3J (the energy level of the laser), 2 mm (laser spot diameter), 5J, 2 mm, and 8J, 2 mm, respectively.

2.2. Validation of Strengthening Parameters for 1Cr18Ni9Ti Materials

Laser-impact-strengthening experiments were performed using the YS0805-R200A laser-impact-strengthening equipment, and the system uses a LAMBER-H08 YAG solid-state laser with the following settings: maximum pulse energy 8J, pulse width 18~20 ns, focused spot 2~5 mm, pure water as a constraint layer, controlling the water flow on the surface of the workpiece to maintain the thickness of 1~3 mm, the absorption of the protective layer of black tape, a spot overlap rate of 50%.

Rotational bending fatigue round bar specimens were clamped and impact-strengthened in the manner shown in Figure 11. Specimens before and after the impact of the are shown in Figure 12.

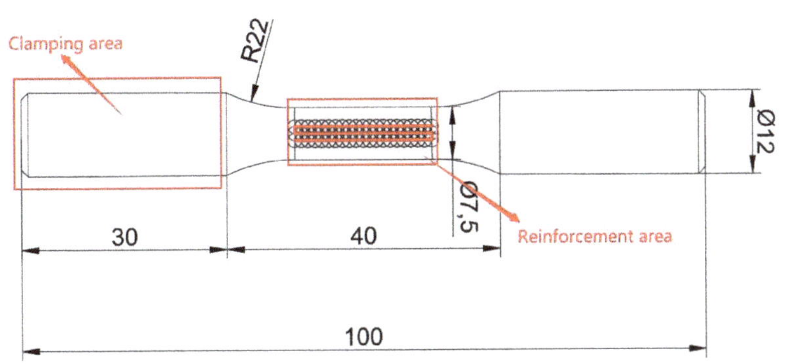

Figure 11. Schematic of clamping and impact areas.

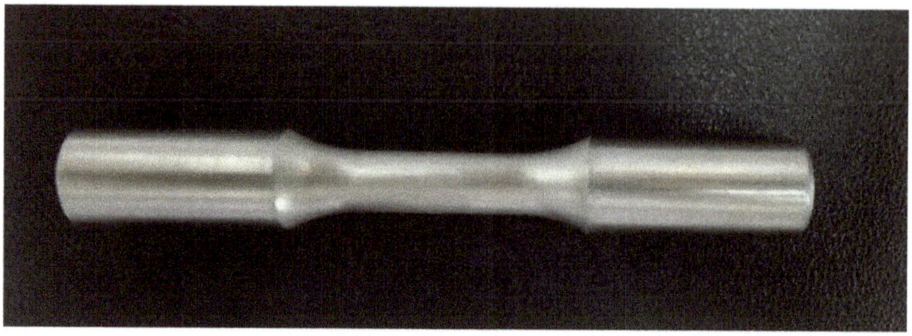

(a)

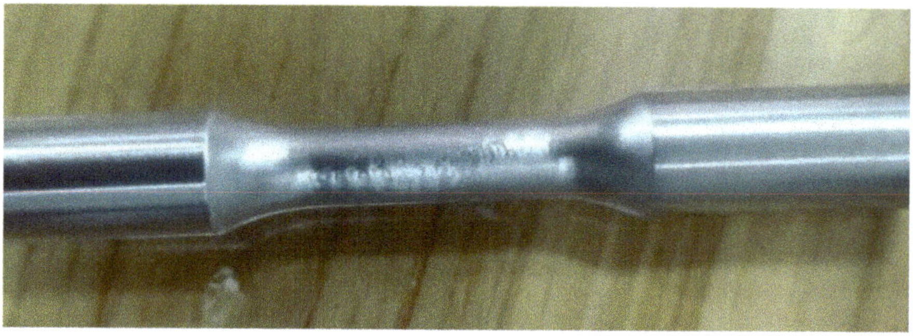

(b)

Figure 12. Comparison of specimens before (**a**) and after (**b**) impact strengthening.

LSP can generate residual compressive stress to inhibit surface crack initiation and expansion [23,24], significantly improving the overall fatigue life of components [25]. Therefore, in this paper, a QBWP-10000 rotary bending fatigue testing machine was used to detect the fatigue life of the test pieces before and after strengthening; the X-ray diffraction method was used to measure the residual compressive stress on the surface of the material; a Vickers hardness tester was used to detect the hardness of the material; a scanning electron microscope (SEM) was used to analyze the fatigue fracture morphology; electron backscattering diffraction (EBSD) was used to observe and analyze the microstructure of the surface layer after strengthening. The microstructure of the strengthened surface layer was observed and analyzed by EBSD.

3. Analysis of Experimental Results

3.1. Analysis of Rotational Bending Fatigue Test Results

The experimental data are shown in Table 4. The experimental final load of 498 MPa was determined by the first eight failures, and the average number of load cycles of the unprocessed original was 258,303 under the same load, while the number of load cycles was significantly increased after LSP, with an average of 369,587 cycles under the 8J process and an average of 439,547 cycles under the 5J process, which is about 1.7-times higher compared to the original fatigue. The 3J and 2 mm processes had the greatest improvement in fatigue performance due to the minimization of the internal tensile stress in the specimen and the smaller influence of the surface morphology, and the number of rotational cycles passed 10^7 times. The average data under different parameters are shown in Figure 13.

Table 4. Rotational bending fatigue test data.

Experimental Temperature (°C)	Maximum Stress (MPa)	Frequency of Tests (Hz)	Stress Ratio R	Number of Cycles (Times)	Results	Process Parameters
23	540	100	−1	7026	Fail	unprocessed
23	500	100	−1	102,080	Fail	
23	470	100	−1	10,000,000	Pass	
23	485	100	−1	10,000,000	Pass	
23	493	100	−1	10,000,000	Pass	
23	498	100	−1	121,356	Fail	
23	498	100	−1	301,411	Fail	
23	498	100	−1	352,114	Fail	
23	498	100	−1	414,107	Fail	8J 2 mm
23	498	100	−1	325,067	Fail	
23	498	100	−1	492,191	Fail	5J 2 mm
23	498	100	−1	386,903	Fail	
23	498	100	−1	10,000,000	Pass	3J 2 mm
23	498	100	−1	10,000,000	Pass	
23	498	100	−1	10,000,000	Pass	
23	498	100	−1	10,000,000	Pass	
23	498	100	−1	10,000,000	Pass	

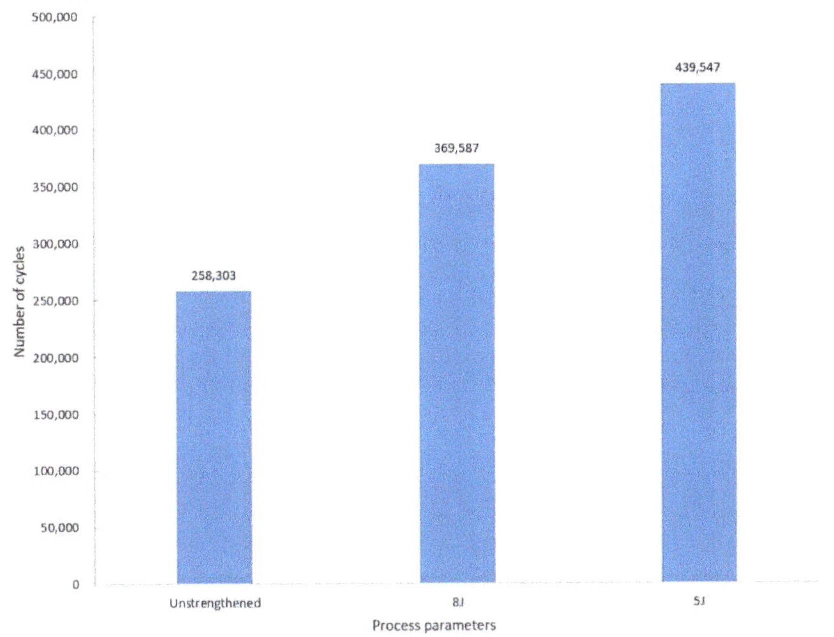

Figure 13. Average number of cycles for different processes.

3.2. Analysis of Hardness Results

The machine was set to pressurize and unload for a time of 12 s, with a holding time of 15 s, and an average of seven positions were measured for each workpiece. The hardness microscope directly measured the diagonal lengths d_1 and d_2 of the indentation area and calculated the hardness (HV). After the measurement was completed, the experimental data were exported through the system, as shown in Figure 14.

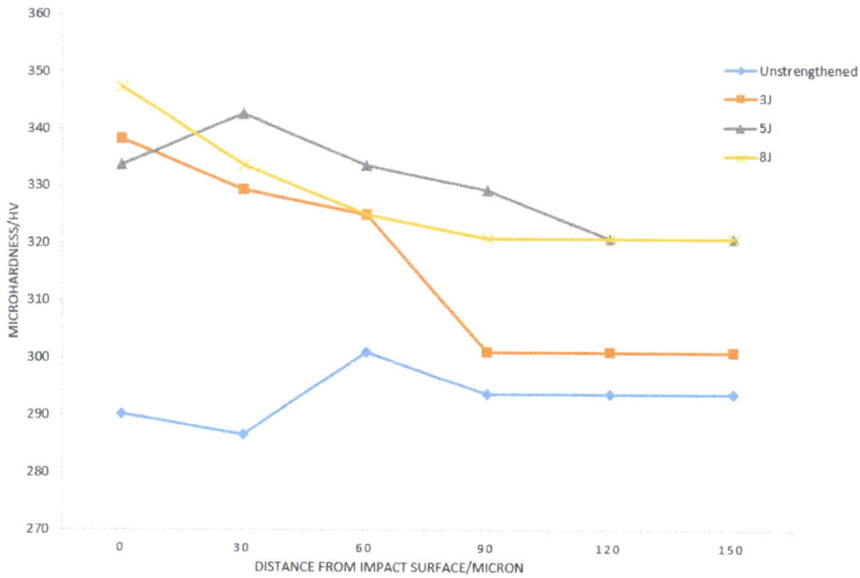

Figure 14. Hardness measurement results.

The experimental results showed that the average hardness of the unprocessed parts was 293.2 HV, and the hardness of the LSP-strengthened parts was significantly improved up to 347.3 HV, while the micro-hardness gradually decreased along the depth direction due to the reflection of the shock wave in the internal part of the material; the results showed that the micro-hardness was maximally increased by 19.3%. This was due to the high-pressure plasma shock wave generated by the laser impact pressure up to several GPa, inducing plastic deformation up to a 10^7 s^{-1} high strain rate on the surface of the metal material [26], generating more dislocations, thus increasing the surface hardness of the material [27].

3.3. Residual Compressive Stress Result on Specimen's Surface

Residual stress on the surface of the material has an important impact on the material's properties, residual compressive stress can inhibit the generation and expansion of cracks on the surface of the material [28,29]. Residual compressive stress will decrease with depth [30]. The results of the laser impact strengthening of the 1Cr18Ni9Ti stainless steel surface's X-direction residual stress are shown in Table 5; the results show that the data obtained under the 3J parameter is the best. The experimental data line chart is shown in Figure 15.

Table 5. Residual compressive stress results at different depths on the specimen surface.

	Specimen 1 (3J, 2 mm)	Specimen 2 (5J, 2 mm)	Specimen 3 (8J, 2 mm)
0 μm (depths)	−464 MPa	−447 MPa	−400 MPa
150 μm	−579 MPa	−530 MPa	−465 MPa
300 μm	−704 MPa	−468 MPa	−429 MPa
450 μm	−439 MPa	−400 MPa	−319 MPa
600 μm	−248 MPa	−355 MPa	−185 MPa

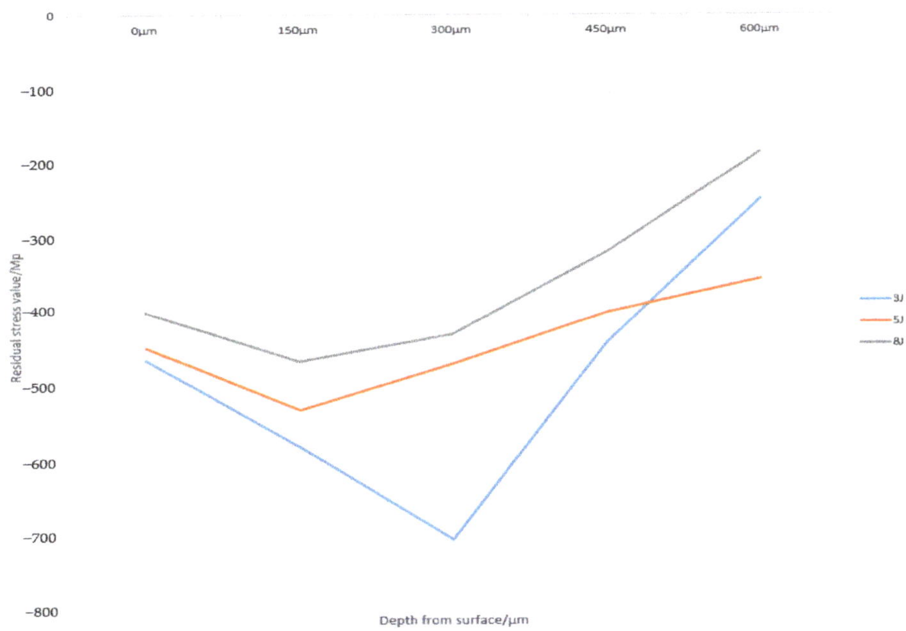

Figure 15. Residual stress line plot.

3.4. Fatigue Specimen Fracture Observation and Analysis

The fracture morphology is shown in Figure 16. The original section can be divided into three areas, A, B, and C. Area A is a relatively straight area of about 1/2 of the section; the cross-section shows that the crack originated from the elliptical marked part of region A, so region A is the crack source area, while region B has a rough and ductile morphology, which is a fatigue transient zone.

Under the 5J and 8J processes, the section can be divided into two regions, A and B, where Region A is relatively straight and has visible crack, which is the crack source region, and Region B is the transient fracture region.

Separate analysis of each specimen's fracture showed that the original crack had a river-like diffraction pattern and a large number of friction–extrusion traces, and the crack was generated from the surface of the specimen and extended inward. Cracks under a 5J impact also appeared near the edges, but did not originate from the surface, which can indicate the inhibition of crack initiation on the surface of the material after impact strengthening. The fatigue life was also improved as a result.

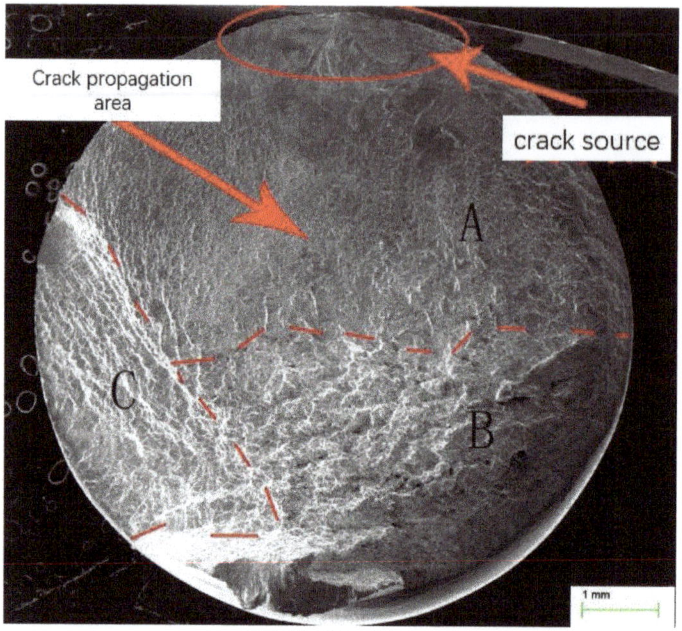

(a)

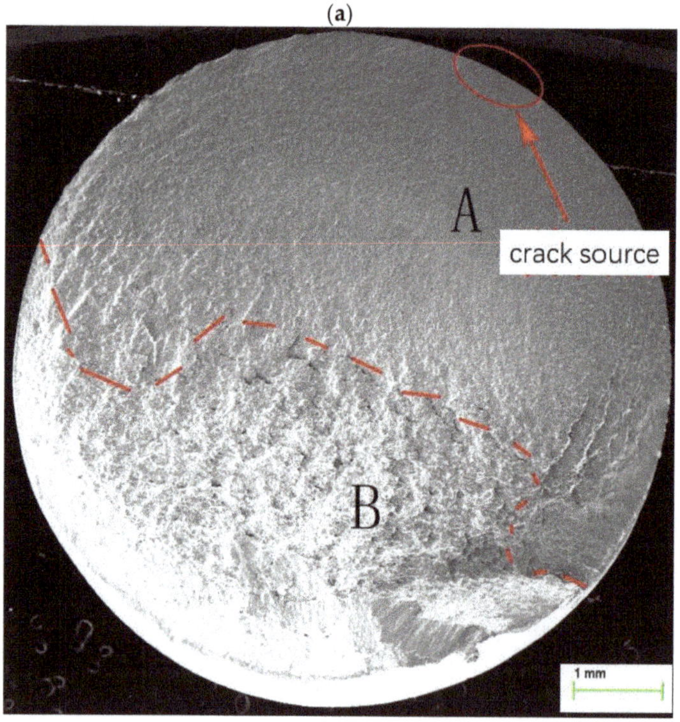

(b)

Figure 16. *Cont.*

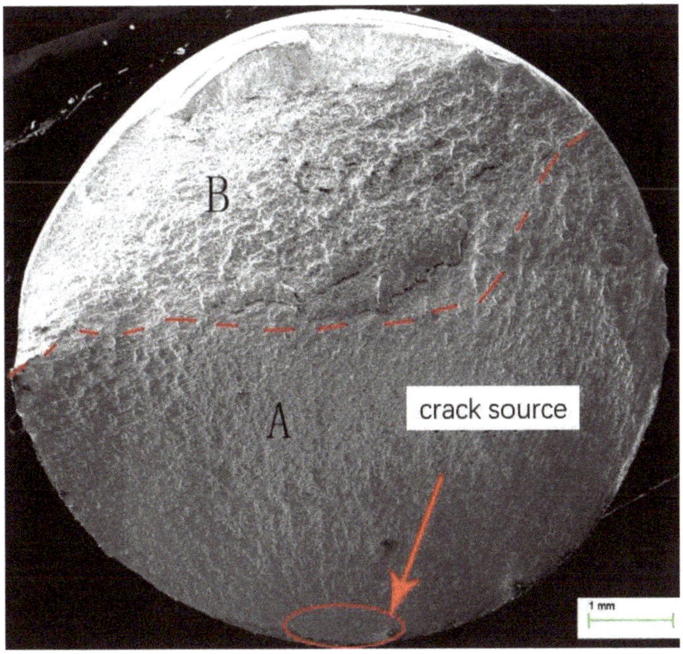

(c)

Figure 16. The macroscopic fracture morphology of unreinforced specimens (**a**), specimens after 5J strengthening (**b**), and specimens after 8J strengthening (**c**).

In Figure 17, by comparing and observing the fatigue transient fracture zones of the three fracture surfaces, it can be seen after magnification that the main morphology is a large and dense number of dimples, which also indicates that they are all ductile fractures. Generally, when the fracture conditions are the same, the larger the size of the dimple, the better the plasticity of the material. As shown in Figure 17, the number and density of the dimples under 5J increased compared with the other two, and the size of the dimples under 8J was obviously larger, so the plasticity of the material was significantly improved by the laser impact strengthening. The increase in plasticity also means that the material is less prone to fracture and the fatigue limit has increased.

3.5. Analysis of EBSD Experiment Results

The test article reverse pole diagram and local orientation difference distribution diagram (KAM) at different process parameters are as follows.

Figure 18 shows that the grain size of the unprocessed specimen was the coarsest and uneven, while the grain size of the intensified specimen was refined to different degrees, and a nanolayer with a depth of about 2 μm was formed on the surface; a high dislocation density near the impacted surface can be clearly seen in the KAM diagram. A comparison of the images shows that the laser energy of 5J produced the largest high-density dislocation area, but the energy of the 8J surface grain refinement was the highest. This also indirectly explains why the fatigue life has been improved [29,31].

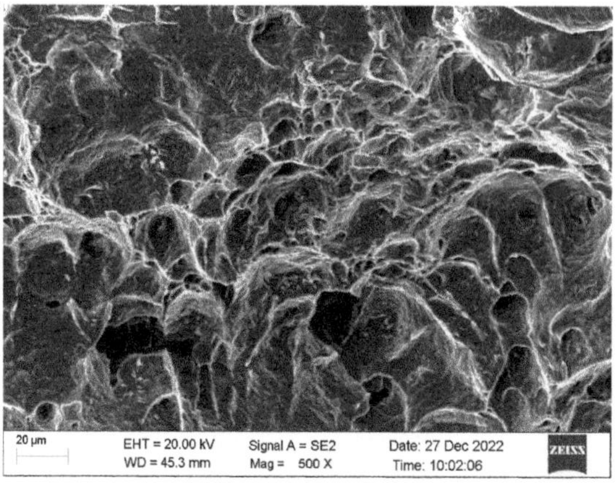

(**a**) Unstrengthened piece

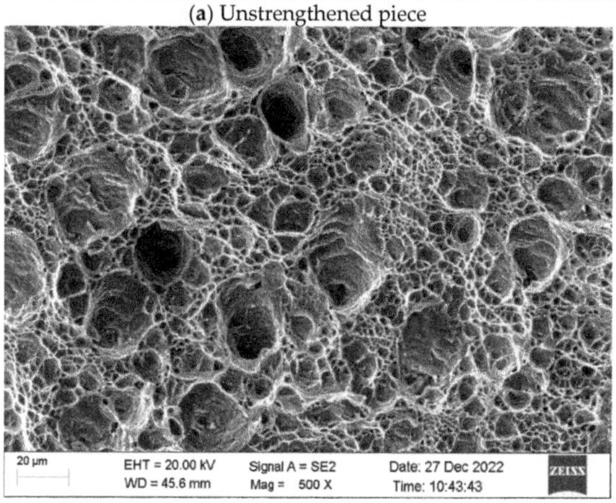

(**b**) 5J

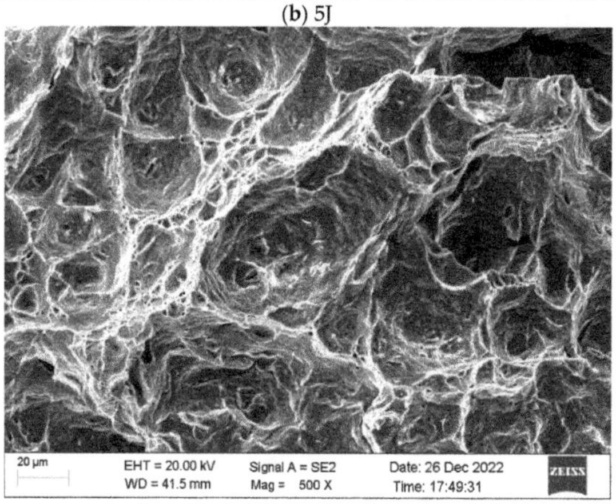

(**c**) 8J

Figure 17. Images of transient fracture fatigue zone under different processes.

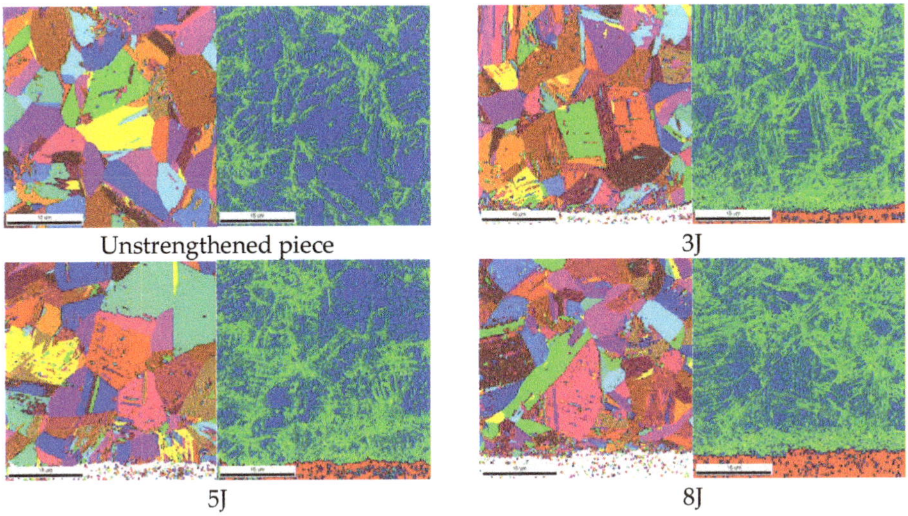

Figure 18. EBSD plot and KAM plot for different processes.

From the size–angle distribution diagram, it can be seen that, after the material undergoes stress deformation, a large number of dislocations are accumulated in the interior, and with the increase of the dislocation density, subcrystalline boundaries are formed inside the material, which further constitutes to the subcrystals. When the stress accumulates to a certain critical value, the subcrystalline boundaries will continue to absorb dislocations, resulting in dynamic recrystallization, and gradually be transformed into large-angle grain boundaries. Therefore, a large number of dislocations exist within the deformed grain, and these dislocations form general deformed grain boundaries with an orientation difference of no more than 2° on both sides, while the angle of the subcrystalline boundaries is generally between 2 and 15° and the angle of the large-angle grain boundaries is more than 15°. From Figure 19a, it can be seen that the proportion of subgrain boundaries with an orientation difference of 2–15° on both sides of the grain boundary on the surface of the raw material is 24.8%. From Figure 19b, it can be seen that the proportion of subgrain boundaries with an orientation difference of 2–15° on both sides of the grain boundary on the surface of the material processed under the 3J parameter is 26.7%. From Figure 19c, it can be seen that the proportion of subgrain boundaries with an orientation difference of 2–15° on both sides of the grain boundary on the surface of the material processed under the 5J parameter is 23.3%. From Figure 19d, it can be seen that the proportion of subgrain boundaries with an orientation difference of 2–15° on both sides of the grain boundary on the surface of the material processed under the 8J parameter is 23.5%. This indicates that the material increases the number of high-density dislocations under the 3J and 5J parameters, while the effect of increasing dislocations under the 8J parameter is poor. This is due to the excessive impact pressure causing the internal plastic wave to rebound and generate tensile stress in the material, resulting in weaker grain refinement and a decrease in the shear dislocation area.

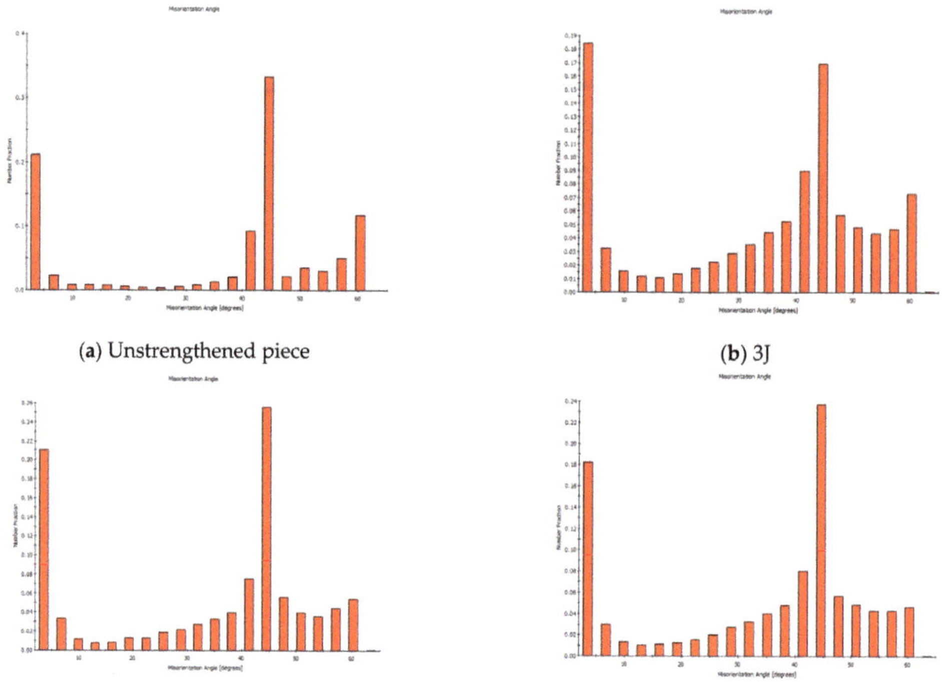

(**a**) Unstrengthened piece (**b**) 3J (**c**) 5J (**d**) 8J

Figure 19. Distribution of large and small angular grain boundaries under different processes.

4. Discussion

This article explored the laser shock peening effect of materials through simulation and then verified the strengthening effect through the fatigue testing of the specimens. The specimens were subjected to hardness testing, residual compressive stress, fracture analysis, and microstructure observation, explaining from multiple perspectives how laser shock peening can improve the fatigue life of specimens. The summary is as follows:

(1) Based on the Hugoniot elastic limit, the simulation parameters were determined to explore the strengthening effect of the specimen. The simulation results showed that the specimen obtained the best strengthening effect under a peak pressure of 3000 MPa.

(2) Fatigue tests were conducted on the specimens, and the fatigue life of the specimens was compared before and after strengthening. The results showed that the fatigue life improvement was the highest under the 3J and 2 mm parameters, which is consistent with the optimal strengthening effect under the corresponding simulation of 3000 MPa.

(3) After laser shock peening, a deep residual compressive stress layer can be generated on the surface of the material, which can effectively suppress the initiation and expansion of surface cracks, and the suppression of cracks can improve the fatigue life. Among them, the best stress effect was generated under the parameters 3J and 2 mm.

(4) Laser shock peening can effectively improve the microhardness of the 1Cr18Ni9Ti's surface, but its strengthening effect weakens with increasing depth. The maximum microhardness can be increased by 19.3% under the 8J process, 17.7% under the 5J process, and 16.2% under the 3J process. The increase in hardness also indicates an increase in fatigue strength.

(5) Scanning electron microscope observation of the fatigue fracture and EBSD experiments found that a high density of dislocations occurred inside the specimen after impact, which was able to delay the expansion rate of fatigue cracks inside the specimen, thus prolonging the fatigue life of the specimen, and the fast fracture zone of the treated specimen showed rough dimples of a relatively large size and depth, thus explaining the reason for the increase of the fatigue life of the material due to the laser-strengthening effect, from a microscopic point of view. The best effect of grain refinement was achieved with the 3J process.

The above experimental results showed that the specimens were optimally strengthened under the parameters 3J and 2 mm, which is consistent with the optimal strengthening effect under the simulated peak pressure of 3000 MPa.

According to the above results, the fatigue performance of the 1Cr18Ni9Ti material was significantly improved by laser impact strengthening, and it can be inferred that laser impact strengthening can improve the fatigue performance of aviation conduits made of the 1Cr18Ni9Ti material.

Author Contributions: X.D.: conceptualization, investigation, writing—original draft; S.M.: methodology, validation; J.Z. (Junlong Zhang): investigation, data curation; Z.J.: methodology; H.L.: investigation; S.W.: validation; C.W.: methodology; J.Z. (Jida Zhong): conceptualization, methodology, writing—review and editing. All authors have read and agreed to the published version of the manuscript.

Funding: This work was supported by the Key Research and Development Program of Jiangxi Province (Grant Number 20223BBE51004).

Data Availability Statement: The data that support the findings of this study are available from the corresponding author upon reasonable request.

Conflicts of Interest: The authors declare no conflict of interest.

References

1. Zhuo, Y.; Chen, Y.; Yang, C. Research status and prospect of welding repair technology for aero-engine blades. *Aerosp. Manuf. Technol.* **2021**, *64*, 22–28.
2. Logesh, M.; Selvabharathi, R.; Thangeeswari, T.; Palani, S. Influence of severe double shot peening on microstructure properties of Ti 6Al-4V and Titanium Grade 2 dissimilar joints using laser beam welding. *Opt. Laser Technol.* **2020**, *123*, 105883. [CrossRef]
3. Leo, P.; D'Ostuni, S.; Nobile, R.; Mele, C.; Tarantino, A.; Casalino, G. Analysis of the process parameters, post-weld heat treatment and peening effects on microstructure and mechanical performance of Ti-Al dissimilar laser weldings. *Metals* **2021**, *11*, 1257. [CrossRef]
4. Chattopadhyay, A.; Muvvala, G.; Sarkar, S.; Racherla, V.; Nath, A.K. Effect of laser shock peening on microstructural, mechanical and corrosion properties of laser beam welded commercially pure titanium. *Opt. Laser Technol.* **2021**, *133*, 106527.
5. Feng, X.; Pan, X.; He, W.; Liu, P.; An, Z.; Zhou, L. Improving high cycle fatigue performance of gas tungsten arc welded Ti6Al4V titanium alloy by warm laser shock peening. *Int. J. Fatigue* **2021**, *149*, 106270. [CrossRef]
6. Shi, X.; Feng, X.; Teng, J.; Zhang, K.; Zhou, L. Effect of laser shock peening on microstructure and fatigue properties of thin-wall welded Ti-6Al-4V alloy. *Vacuum* **2021**, *184*, 109986. [CrossRef]
7. Trdan, U.; Skarba, M.; Grum, J. Laser shock peening effect on the dislocation transitions and grain refinement of Al–Mg–Si alloy. *Mater. Charact.* **2014**, *97*, 57–68.
8. Peyre, P.; Carboni, C.; Forget, P.; Beranger, G.; Lemaitre, C.; Stuart, D. Influence of thermal and mechanical surface modifications induced by laser shock processing on the initiation of corrosion pits in 316L stainless steel. *J. Mater. Sci.* **2007**, *42*, 6866–6877. [CrossRef]
9. Irizalp, S.G.; Saklakoglu, N.; Akman, E.; Demir, A. Pulsed Nd:YAG laser shock processing effects on mechanical properties of 6061-T6 alloy. *Opt. Laser Technol.* **2014**, *56*, 273–277. [CrossRef]
10. Gao, Y.; Yang, W.; Huang, Z.; Lu, Z. Effects of residual stress and surface roughness on the fatigue life of nickel aluminium bronze alloy under laser shock peening. *Eng. Fract. Mech.* **2021**, *244*, 107524.
11. Meng, L.; Zhao, S.; Liang, Z.; Lv, Y. Study on the Influence of Surface Roughness and Roughness Ratio on the Fatigue Performance of DD10 Alloy. *Shanxi Metall.* **2023**, *46*, 4–6+9.
12. Fueki, R.; Takahashi, K. Improving the fatigue limit and rendering a defect harmless by laser peening for a high strength steel welded joint. *Opt. Laser Technol.* **2021**, *134*, 106605. [CrossRef]
13. Soyama, H. Laser Cavitation Peening and Its Application for Improving the Fatigue Strength of Welded Parts. *Metals* **2021**, *11*, 531. [CrossRef]

14. Soyama, H.; Simoncini, M.; Cabibbo, M. Effect of Cavitation Peening on Fatigue Properties in Friction Stir Welded Aluminum Alloy AA5754. *Metals* **2021**, *11*, 59. [CrossRef]
15. Yu, K.; Wang, L.; Zhao, T.; Shu, S.; Zhou, L. Optimization of Residual Stress Field and Improvement of Fatigue Properties of Thin-Walled Pipes by Filling Laser Shock Peening. *Metals* **2022**, *12*, 1733. [CrossRef]
16. Jiang, F.; Li, J.; Sun, J.; Zhang, S.; Wang, Z.; Yan, L. Al7050-T7451 turning simulation based on the modified power law material model. *Int. J. Adv. Manuf. Technol.* **2010**, *48*, 871–880. [CrossRef]
17. Steinberg, D.J.; Cochran, S.G.; Guinan, M.W. A constitutive model for metals applicable at high rate. *J. Appl. Phys.* **1980**, *51*, 1498–1504. [CrossRef]
18. Zerilli, F.J.; Armstrong, R.W. Dislocation-mechanics-based constitutive relations for materials dynamic calculations. *J. Appl. Phys.* **1987**, *61*, 1816–1825. [CrossRef]
19. Johnson, G.R.; Cook, W.H. A constitutive model and data for metals subjected to large strains, high strain rates and high temperatures. In Proceedings of the 7th International Symposium on Ballistics, Hague, The Netherlands, 19–21 April 1983; pp. 541–547.
20. Wu, J. Basic Research on Drilling Technology of Typical Difficult-to-Machine Materials. Ph.D. Thesis, Harbin Institute of Technology, Harbin, China, 2010; pp. 64–78.
21. Chou, P.C.; Hopkins, A.K. *Dynamic Response of Materials to Intense Impulsive Loading*; Defense Technical Information Center: Belvoir, VA, USA, 1973.
22. Ballard, P.; Fournier, J.; Fabbro, R.; Frelat, J. Residual stresses induced by laser-shocks. *Le Journal de Physique IV* **1991**, *1*, C3-487–C3-494. [CrossRef]
23. Montross, C.S.; Wei, T.; Ye, L.; Clark, G.; Mai, Y.W. Laser shock processing and its effects on microstructure and properties of metal alloys: A review. *Int. J. Fatigue* **2002**, *24*, 1021–1036. [CrossRef]
24. Ruschau, J.J.; John, R.; Thompson, S.R.; Nicholas, T. Fatigue crack nucleation and growth rate behavior of laser shock peened titanium. *Int. J. Fatigue* **1999**, *21*, S199–S209. [CrossRef]
25. Zhang, Y.; You, J.; Lu, J.; Cui, C.; Jiang, Y.; Ren, X. Effects of laser shock processing on stress corrosion cracking susceptibility of AZ31B magnesium alloy. *Surf. Coat. Technol.* **2010**, *204*, 3947–3953. [CrossRef]
26. Li, Y. *Theory and Technology of Laser Impact Peening*; Science Press: Beijing, China, 2013.
27. Nie, X.; Long, N.; Liu, H.; He, W.; Li, Q. Effect of laser impact strengthening on surface hardness of TC17. *Mech. Des. Manuf.* **2012**, 198–200. [CrossRef]
28. Xiong, J.; Mao, W.; Yu, T. Effect of residual stress on fatigue crack extension. *J. Hubei Inst. Technol.* **1996**, *4*, 9–13.
29. Zhou, L.; Zhao, T.; Yu, Y.; Liu, P.; Pan, X. Effect of Laser Shock Peening on High-Cycle Fatigue Performance of 1Cr18Ni9Ti/GH1140 Weld. *Metals* **2022**, *12*, 1495. [CrossRef]
30. Sanchez, A.G.; You, C.; Leering, M.; Glaser, D.; Furfari, D.; Fitzpatrick, M.E.; Wharton, J.; Reed, P.A.S. Effects of laser shock peening on the mechanisms of fatigue short crack initiation and propagation of AA7075-T651. *Int. J. Fatigue* **2021**, *143*, 106025. [CrossRef]
31. Zhang, W.; Simpson, C.A.; Lopez-Crespo, P.; Mokhtarishirzabad, M.; Buslaps, T.; Pippan, R.; Withers, P.J. The effect of grain size on the fatigue overload behaviour of nickel. *Mater. Des.* **2020**, *189*, 108526. [CrossRef]

crystals MDPI

Article

Improvement of Hydrogenation and Dehydrogenation Kinetics of As-Cast AZ91 Magnesium Alloy via Twin Parallel Channel Angular Extrusion Processing

Mohammad Abdi [1], Ramin Ebrahimi [1,*] and Ebad Bagherpour [2]

[1] Department of Materials Science and Engineering, School of Engineering, Shiraz University, Shiraz 71348-51154, Iran
[2] Brunel Centre for Advanced Solidification Technology (BCAST), Brunel University London, Uxbridge, Middlesex UB8 3PH, UK
* Correspondence: ebrahimy@shirazu.ac.ir

Abstract: In the current study, Twin Parallel Channel Angular Extrusion (TPCAE) as a developed SPD processing technique is used to improve the hydrogen storage properties of AZ91 cast alloy. The processing is conducted at different temperatures, ranging from 340 °C down to 200 °C. The hydrogen absorption and desorption tests are conducted kinetically at three different temperatures, using a Sievert-type apparatus. Remarkable improvement in the absorption kinetic is achieved as a result of the TPCAE processing. A maximum absorption capacity of 6.1 wt.% within a time span of 2000 s is achieved for the sample with three passes of processing complemented at 250 °C. Also, the kinetic of dehydrogenation is improved significantly and complete desorption at 350 °C is achieved for all the processed samples within a time span of maximum 2500 s. By calculating the activation energy of hydrogenation and evaluating the microstructure changes, it is found that implementing sufficient thermomechanical work level along with applying the last pass of the process at lower temperature results in a reduction of the activation energy and improvement of the hydrogenation kinetic.

Keywords: AZ91 magnesium alloy; severe plastic deformation; hydrogen storage; kinetic

1. Introduction

Nowadays, fossil fuels have caused significant problems, including global warming and climate change. Hydrogen can be considered as an alternative energy resource, especially in the automotive industry and future energy systems due to its safety, abundance, and cleanness [1]. However, extensive use of hydrogen is hindered by its storage difficulties. Storage in the solid state via metal hydrides presents advantageous features, such as high gravity storage capacity and safety [2]. Mg is an attractive material in new research areas such as the ion battery, hydrogen storage, and biomaterials due to the lightness, recyclability, corrosion resistance, and low cost [3]. In the case of hydrogen storage, Mg has a high gravimetric capacity (7.6 wt.%) and has the potentiality to be used for onboard vehicular applications [1]. Nonetheless, there are major problems with the vast mobile application of MgH_2, including poor kinetics of absorption/desorption of hydrogen and high stability of MgH_2, leading to increase its dissociation temperature [1]. The poor kinetics are mainly due to the low diffusivity of hydrogen in the Mg crystalline lattice and the presence of thick surface oxide layer hindering H_2 dissociation and penetration of H atoms into the bulk [4].

In this regard, research works have been performed to improve the hydrogenation performance of Mg. From the thermodynamic viewpoint, alloying with other elements, such as Al, decreases the hydrogen binding energy [5,6]. Although Al does not participate in the hydrogen absorption, it can destabilize MgH_2 and improve desorption efficiency [7]. Catalytic additions (Pd, Fe, V, Ni, Ti, and Nb_2O_5) to the Mg matrix have been reported to enhance H_2 dissociation and improve hydrogenation kinetic [8]. Moreover, the kinetic of hydrogen storage can be improved significantly by refining the microstructure and producing a high fraction of structural defects, such as vacancies, dislocations, and grain boundaries with low activation energy of diffusion [9]. These defects, especially high angle grain boundaries, can act as straightforward paths for diffusion and absorption of hydrogen atoms into the Mg bulk [10].

High energy ball milling (HEBM) [1,11] is a conventional method used vastly to make nanostructure materials consisting of ultrafine grains and lattice defects, which can significantly improve the hydrogenation kinetic. However, HEBM has some limitations, including high contamination, intensive air reactivity causing fire hazard, and high energy consuming and/or high cost for production of small quantities [12]. From this perspective, these limitations would be addressed by bulk processing using severe plastic deformation (SPD), in which intensive strain is accumulated by repetition of plastic deformation [13]. By applying SPD, ultrafine grain (UFG) structures are obtained in favor of the diffusion of hydrogen atoms, which in turn avoids contamination and fire risk.

Mg and its alloys are of limited workability due to the lack of sufficient active slip systems in the hexagonal structure and their SPD processing is challenging. Nonetheless, some strategies can be considered to overcome these challenges. Some of these include applying SPD at high temperatures to activate the pyramidal slip systems, increasing the die angles to reduce the strain localized points, and using backward pressure [14]. Also, some other works such as prestraining or developing a texture have been conducted to improve the formability and mechanical behavior of Mg alloys [15]. Recently, investigations were conducted to study the ability of various SPD techniques, including equal channel angular extrusion (ECAE) [7,16–18], high-pressure torsion (HPT) [19,20], and accumulative roll bonding (ARB) [21] in improvement of the hydrogen storage properties of Mg-based alloys. Skripnyuk et al. pioneered the use of ECAE for the purpose [22,23]. They also investigated a combination of the ECAE and HEBM processes to improve the kinetic properties of the ZK60 alloy and achieved the absorption capacity of 6.2 wt.% at 300 °C [23]. Asselli et al. [24] used both ECAE and ARB to improve the hydrogenation behavior of ZK60 alloy, on the increase of the maximum storage capacity from 4.77 wt.% (by only ECAE processing) up to 6.4 wt.% (by combination of ECAE and ARB processing). Botta et al. [25] investigated the processing effect of HPT and cold rolling, separately, to improve the absorption properties of bulk Mg samples. They were also successful in increase of maximum absorption capacity of pure Mg ribbons produced by melt spinning for almost 65% due to the formation of a strong (0001) texture during cold rolling. Texture effects, formed during rolling, on the improvement of hydrogenation kinetics of pure Mg and its alloys, AZ91 and AM60D, were also reported in other studies [21,26]. Furthermore, Huot et al. [27] have reported the different effects of ECAE processing routes in improving hydrogenation kinetics of Mg alloys. In this regard, using processing route of A (in which the sample is pressed repetitively without rotation) in ECAE delivers a very marked texture for AZ31 Mg alloy, while using route Bc (in which the sample is rotated by 90° in the same sense between each pass) prepares more grain refinements with random texture. Both processing routes deliver samples with the ability to uptake hydrogen up to 7.6 wt.%. However, the samples processed by route A show an incubation time before start of hydrogenation, while the ones processed by route B_c start the hydrogenation readily without any incubation time.

The main drawback of SPD processing is that they have not been industrialized so far. The current study addresses the feasibility of using a novel developed SPD technique, named twin parallel channel angular extrusion (TPCAE), which is one step closer to be industrialized in comparison to the conventional SPD ones, especially in the hydrogen storage application. TPCAE is a developed technique of ECAE in parallel channels, in which less energy per volume of material is consumed, while the tool has much more stability [28]. As cast AZ91 alloy was selected as the hydrogen storage material due to the existence of Al which decreases the hydrogen binding energy and destabilizes the MgH_2 hydride phase. Considering a solution treatment and quenching in water as a preliminary step, TPCAE at a temperature higher than 300 °C was started and continued for several passes by a step-by-step reduction of temperature to 200 °C in different ways. Then, hydrogenation kinetics were measured for different processing conditions. Moreover, the effect of processing route and thermomechanical work level on the hydrogen storage properties of AZ91 alloy were evaluated considering the activation energy concept.

2. Materials and Methods

For TPCAE processing a commercial cast AZ91 Mg alloy was used with the chemical composition illustrated in Table 1.

Table 1. Chemical composition of the AZ91 alloy in wt.%.

Mg	Al	Zn	Mn	Fe	Si	Cu	Ca
Base	8.9	0.85	0.338	0.01	0.08	0.004	0.04

Prior to the processing, a solution heat treatment consisting of homogenization at 430 °C for 10 h and quenching in water was conducted. Afterward, the TPCAE processing in six different configurations, summarized in Table 2, was conducted.

Table 2. Different processing route for SPD of AZ91 alloy by TPCAE method.

Sample	Pass 1	Pass 2	Pass 3	Pass 4
S1	340 °C	-	-	-
S2	340 °C	300 °C	-	-
S3	340 °C	250 °C	-	-
S4	340 °C	200 °C	-	-
S5	340 °C	300 °C	250 °C	-
S6	340 °C	300 °C	250 °C	200 °C

Dislocation density was evaluated based on the modified Williamson–Hall approach, using X-ray diffraction (XRD) patterns of the processed samples. Mean grain size and volume fraction of precipitates were calculated using ImageJ software [29]. More detail on the material, the processing, and the calculations can be found in Ref. [30], published by the current authors.

One of the key factors to improve the hydrogenation properties of Mg alloys is to enhance the area-to-mass ratio of the material to facilitate the H_2 dissociation and penetration of H atoms into the bulk of material [31]. As a result, powders of the processed samples were needed for the investigations. Therefore, the samples were made into powder using a filing process under controlled argon atmosphere to prevent the formation of the thick surface-oxide layer on the powder surface. The activation process was conducted with the following details:

- For each test, 0.22 g powder was placed in the reactor and washed with hydrogen gas, followed by vacuuming to 10^{-2} mbar. The washing process was repeated three times.
- Afterward, it was placed under a highly enhanced vacuum condition with a pressure of 10^{-5} mbar. Then, the temperature was increased up to 150 °C. As the temperature increased, followed by the increase in the pressure, the vacuum pump worked for one hour to achieve the pressure of 10^{-5} mbar at 150 °C.
- Similar procedure was repeated as the temperature increased from 150 °C to 250 °C and from 250 °C to 350 °C, respectively, and the pressure was maintained at 10^{-5} mbar at the end of each temperature-rise step.
- Afterwards, hydrogen gas was injected into the sample at 350 °C under 30 bar pressure, and kept for 15 h.
- In the last stage, the sample powder was placed again under a vacuum of 10^{-5} mbar at 350 °C.

After the activation process, the kinetic measurement of absorption was started at 350 °C by the injection of hydrogen gas at 30 bar following by the desorption measurement at a pressure of 10^{-1} mbar. To perform the test at 300 °C, there was a need to ensure that all the hydrogen atoms are brought to the gas phase. Thus, the vacuum pump worked to lower the pressure down to 10^{-5} mbar, while the reactor temperature was set to 300 °C. Then, the absorption/desorption measurements were repeated at this temperature in a similar condition to that of 350 °C. The same procedure was repeated for the temperature of 250 °C. Hydrogenation and dehydrogenation measurements were conducted according to the same procedure for all samples obtained through the different TPCAE processing conditions, as specified in Table 2. The analysis of XRD was performed after hydrogenation and dehydrogenation, using Cu-Kα peak with the wavelength of 0.154056 nm, a time scale of 1 s, a length scale of 0.05° and an angle range of 20–100°.

3. Results

3.1. Microstructure Evolution

Thermomechanical processing of the solution heat-treated AZ91 alloy results in a remarkable microstructure evolution. Table 3 illustrates the variation of mean grain size, dislocation density, and volume fraction of precipitate (β-$Mg_{17}Al_{12}$ phase) during different conditions of processing. The determinations can be found elsewhere [30].

Table 3. Microstructure evolution for the TPCAE processed samples of AZ91 alloy via different processing routes [30].

Sample	Mean Grain Size (μm)	Dislocation Density ($\times 10^{14}/m^2$)	Volume Fraction of β-Phase Precipitate (%)
As cast	550	0.1 [32]	-
S1	13.54	2.73	3
S2	4.29	1.33	8.7
S3	2.39	0.36	13.3
S4	5.67	4.1	15.3
S5	2.19	0.61	15.4
S6	3.66	3.28	15.6

The primary as cast grain size is reduced via one pass of processing at 340 °C. It should be mentioned that recrystallization and precipitation happen simultaneously during plastic deformation [30]. Precipitation occurs during the processing, owing to the fact that the β phase was eliminated through the solution treatment before the processing, resulting in a supersaturated microstructure. Therefore, there is a potential for precipitation of β phase through the TPCAE processing [33].

By lowering the temperature of the second pass down to 300 °C (S2 sample) and 250 °C (S3 sample), more grain refinement is achieved and precipitation of β phase is enhanced, while the dislocation density is reduced, indicating the dominance of recrystallization. Moreover, at lower temperature the diffusion rate decreases, leading to the prevention of grain growth and retard of precipitate coarsening. Thereby, the fine and dispersed precipitates prevent growth of new recrystallized grains via the pinning effect mechanism [34]. By reducing temperature of the second pass to 200 °C (S4 sample), recrystallization is replaced by twinning, which occurs in the un-recrystallized grains remaining from the first pass of processing at 340 °C [35]. In this condition, the dislocation density increases by one order of magnitude, as compared to the S3 sample.

By increasing level of the thermomechanical processing applied through the step-by-step reduction of the processing temperature, the mean grain size decreases to its lowest size of 2.19 μm (S5 sample). Besides, the volume fraction of β-phase precipitate increases in comparison with the S3 sample and reaches the level of 15.4%. The observed increase of β phase by increasing the accumulative strain is in accordance with the previous report of Huang et al. [7], on the improvement of the hydrogenation behavior of AZ61 alloy by ECAE.

Interestingly, by comparing S5 and S6 samples, an increase in the grain size is observed. The reason is the coarsening of the of the small precipitates due to the further thermomechanical processing which in turn, would lead to the grain coarsening because of the slight weakening of the pinning effect mechanism [33]. On the other hand, the fraction of β-phase precipitate remains almost constant in S4, S5, and S6 samples, indicating that the subsequent thermomechanical processing does not have any significant effect on the fraction of the precipitates after the second pass of the TPCAE; instead, more dispersion of this phase occurs, resulting in a more uniform distributed microstructure [30,33].

3.2. Hydrogen Absorption and Desorption Properties

The absorption and desorption kinetic measurements at 350 °C, 300 °C, and 250 °C are shown for the cast sample in Figure 1. The maximum absorption capacity is about 2.5 wt.%. The hydrogenation kinetic becomes slower by decreasing the temperature. The dehydrogenation is incomplete at all temperatures. However, dehydrogenation kinetic at 350 °C is significantly better, as compared to the lower temperatures.

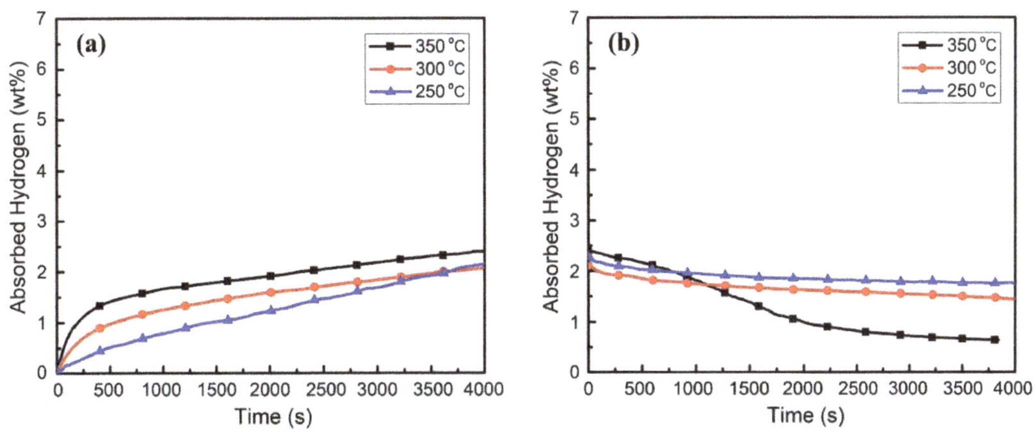

Figure 1. Kinetic of hydrogenation and dehydrogenation for as cast AZ91 alloy at different temperatures: (**a**) Hydrogenation at pressure of 30 bar; (**b**) Dehydrogenation at pressure of 10^{-1} mbar.

Hydrogenation and dehydrogenation behaviors for the S1 and S2 samples are illustrated in Figure 2. As shown in Figure 2a, a single pass of the TPCAE improves the maximum absorption capacity at 350 °C significantly from 2.5% to 4.8 wt.%, as well as a significant enhancement in the hydrogenation kinetics. In addition, for the S1 sample, the hydrogenation trend at 300 °C tends to get considerably close to that at 350 °C, but the kinetic is not as fast as that. Thermomechanical processing does not improve the absorption trend at 250 °C. From Figure 2b it is obvious that a single pass of TPCAE was enough to ensure the complete desorption at 350 °C with a suitable kinetic. However, the dehydrogenation kinetics are significantly slower at lower temperatures. The improvements in absorption could be associated with the considerably lower grain size and the higher dislocation density of the sample after a pass of TPCAE at 340 °C.

After the second pass of TPCAE at 300 °C, the maximum absorption capacity reaches the level of 5.5 wt.% at 350 °C (see Figure 2c). The maximum absorption decreases to about 4.6 wt.% by decreasing the hydrogenation temperature to 300 °C. However, the kinetic of absorption within the first 250 s is higher at 300 °C. Furthermore, a second pass of TPCAE improves the absorption behavior at 250 °C, it can be realized by comparing the absorption behavior of S1 and S2 samples in Figure 2a,c, respectively. The desorption behavior of the S2 sample has a similar trend and kinetic to the S1 (Figure 2d), so that a complete desorption is achieved at 350 °C, while for the lower temperature the kinetics are still very poor.

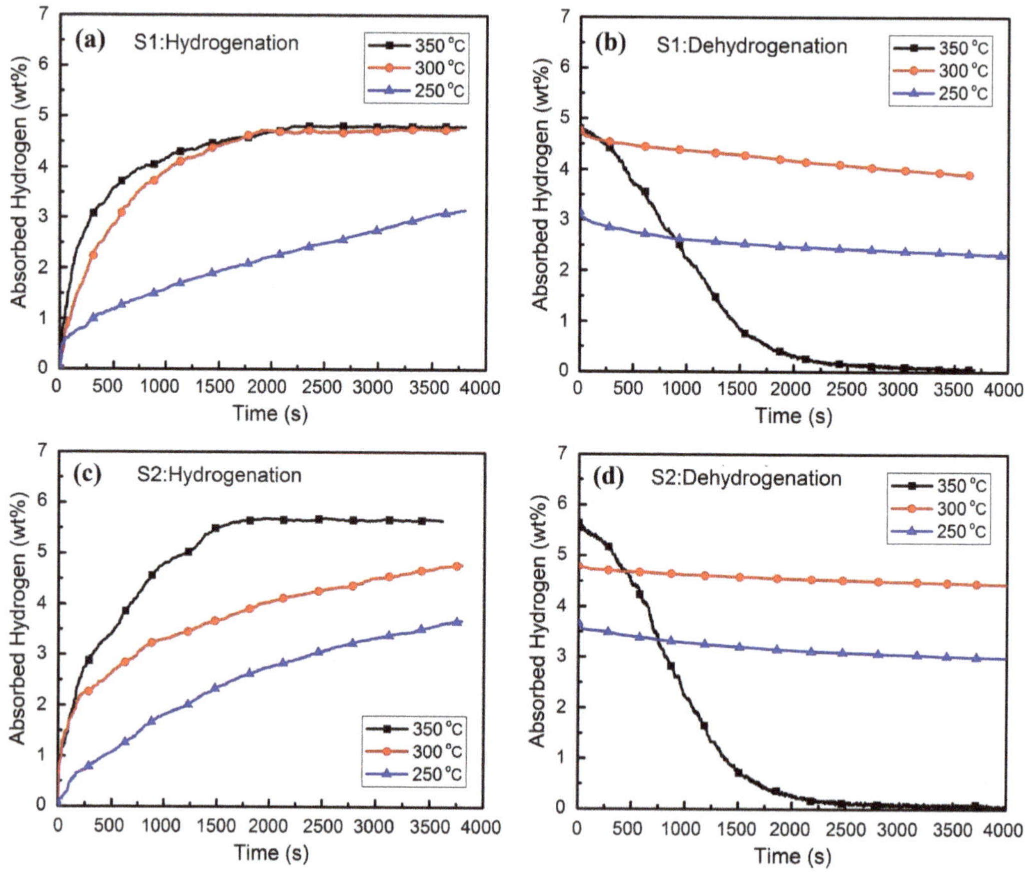

Figure 2. Hydrogenation kinetics at 30 bar and dehydrogenation kinetics at 10^{-1} mbar for different temperatures. (**a**,**b**) S1 sample (a pass of TPCAE at 340 °C), (**c**,**d**) S2 sample (second pass of TPCAE at 300 °C).

Hydrogenation and dehydrogenation kinetics for the S3 and S4 samples are illustrated in Figure 3. Clearly, lowering the second pass processing temperature from 300 °C to 250 °C, makes a huge improvement in hydrogenation by equalizing the hydrogenation trend for all the three temperatures while still the maximum absorption of 5.5 wt.% is achieved at 350 °C. Whereas the maximum absorption at 250 °C for the S2 sample was 3.4 wt.%, it is 4.3 wt.% for the S3 sample, which shows almost 27% increase in the maximum absorption for the lowest hydrogenation temperature. Considering Figure 3b, the dehydrogenation of S3 sample at 350 °C is fully complemented below 1000 s which is much shorter time in comparison with S1 and S2 samples. Comparing the dehydrogenation kinetic of S2 and S3 samples (see Figures 2d and 3b), it is clear that the dehydrogenation kinetic at 300 °C is improved slightly in S3 sample, while at 250 °C no significant improvement is observed. The improvement of hydrogenation/dehydrogenation behavior of the sample processed by two passes of TPCAE in lower temperature is attributed to the higher level of recrystallization occurring at the lower temperatures during the processing, which leads to the formation of a highly refined microstructure with higher fraction of sub-structures and grain boundaries.

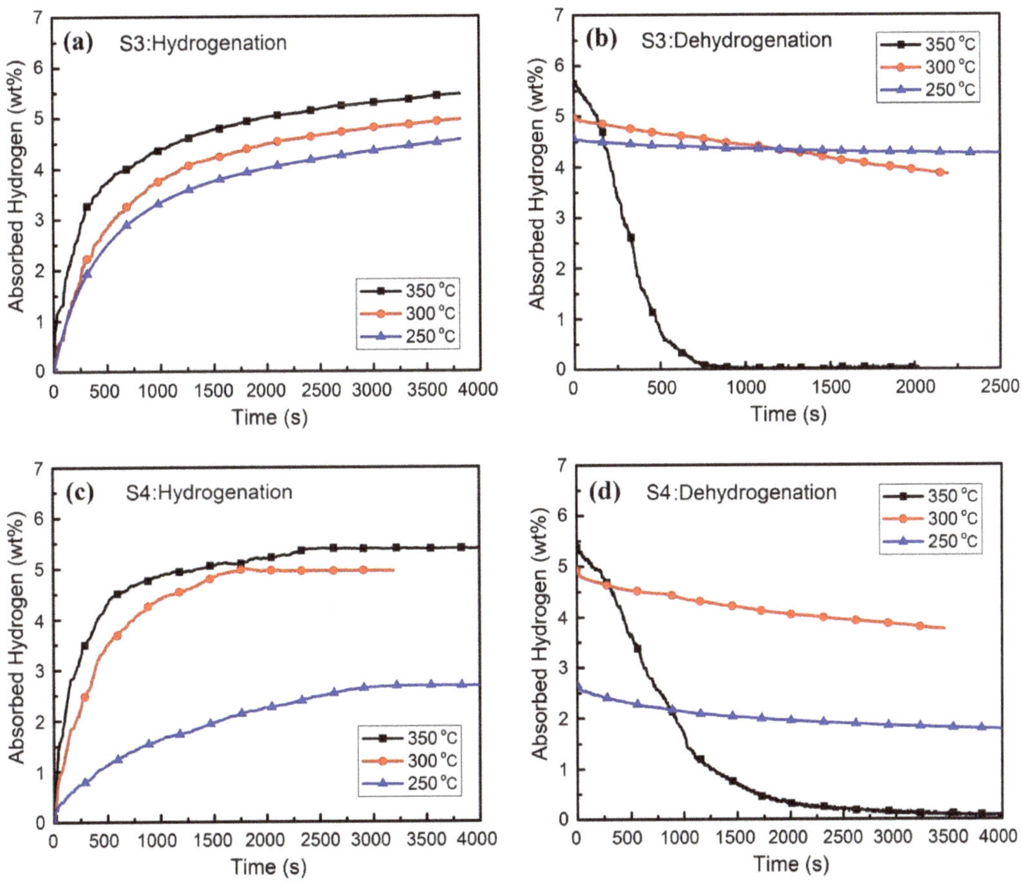

Figure 3. Hydrogenation kinetics at 30 bar and dehydrogenation kinetics at 10^{-1} mbar for different temperatures: (**a**,**b**) S3 sample (second pass of TPCAE at 250 °C); (**c**,**d**) S4 sample (second pass of TPCAE at 200 °C).

Further reduction in the TPCAE processing temperature to 200 °C (S4 sample), does not have any effect on the maximum absorption capacity at 350 °C (see Figure 3c) and it is almost same as what is achieved for S3 sample. In addition, the hydrogenation trend of S4 sample at 300 °C is comparable to that of 350 °C. Nevertheless, unlike S3 sample, the hydrogenation trend of S4 sample at 250 °C suffers from a considerable drop. Comparing the dehydrogenation of S4 sample at 350 °C (Figure 3d) with S3 sample (Figure 3b) shows a considerable increase in the complementation time. The weakening of hydrogenation and dehydrogenation for the sample having a lower TPCAE processing temperature can be described by the existence of a bimodal microstructure [30] originated from activation of twinning mechanism instead of recrystallization in the second pass of the processing at lower temperatures. It is concluded that, for the second pass of the TPCAE, processing at 250 °C (in which recrystallization is the dominant mechanism) results in the best absorption and desorption behavior.

To investigate the effect of hydrogenation and dehydrogenation on the phase transformation, XRD patterns of S4 after hydrogenation and dehydrogenation at 350 °C are illustrated in Figure 4, as examples. New diffraction peaks corresponding to MgH_2 phase are observed after hydrogenation, indicating the formation of a hydride phase. Also, peaks of aluminum appear after hydrogenation, as a result of transformation of the $Mg_{17}Al_{12}$ phase to Mg_2Al_3 and MgH_2, following by a transformation of Mg_2Al_3 to MgH_2 and Al [36,37]. Since still small Mg peaks exist, it shows the hydrogenation process is not completed (Figure 4a). On the other hand, as seen in Figure 4b, almost all of MgH_2 peaks disappear after desorption that indicates the approximately complete dehydrogenation at this temperature. Because of the reverse reactions, including the transformation of Al and MgH_2 to Mg_2Al_3, and then, Mg_2Al_3 and MgH_2 to $Mg_{17}Al_{12}$ phase, there are no Al peaks after dehydrogenation.

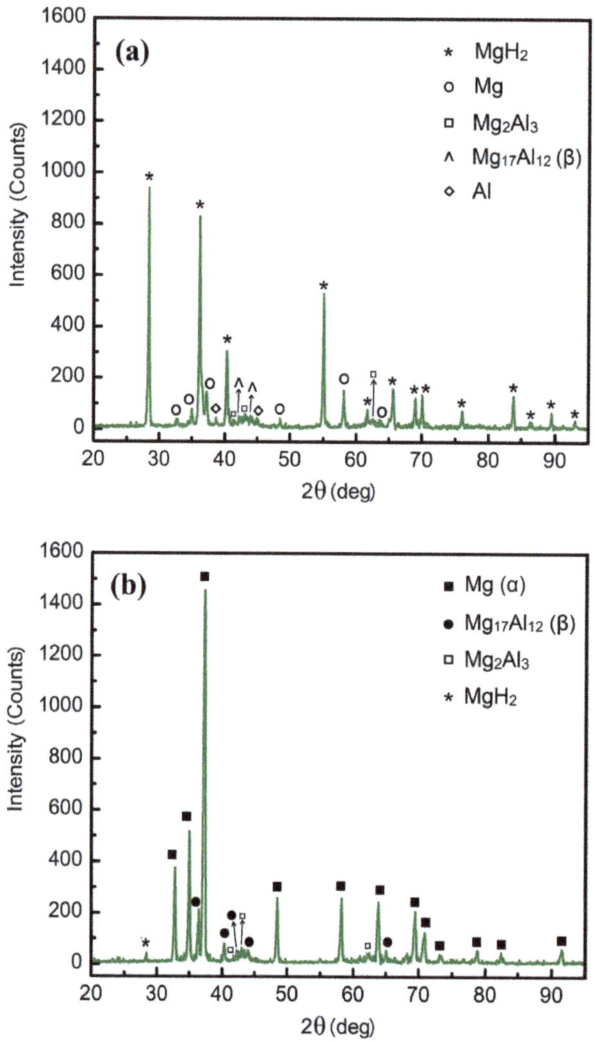

Figure 4. XRD diffraction patterns for S4 sample (second pass of TPCAE at 200 °C): (**a**) after hydrogenation; (**b**) after dehydrogenation.

The hydrogenation and dehydrogenation behaviors for the samples processed by three and four passes of TPCAE are illustrated in Figure 5. After the third pass of TPCAE, the maximum absorption capacity reaches its highest value of 6.1 wt.% (Figure 5a). Looking at the existing literature, reaching a level of 6.1 wt.% absorption capacity in this alloy using only a single SPD method is a significant achievement. For example, in 2017, Jorge et al. [21], has achieved a much lower amount of 4.5 wt.% by means of a combination of the ECAE and ARB processes without sample powdering. In another research, Floriano et al. [38] have managed to increase the hydrogen absorption capacity to 5.5 wt.% by performing low temperature rolling. The closest achieved maximum absorption capacity to the current research, has been reported by Amira et al. [39] in 2012, where the same level of hydrogen absorption capacity (6.1 wt.%) has been attained by a high energy ball milling.

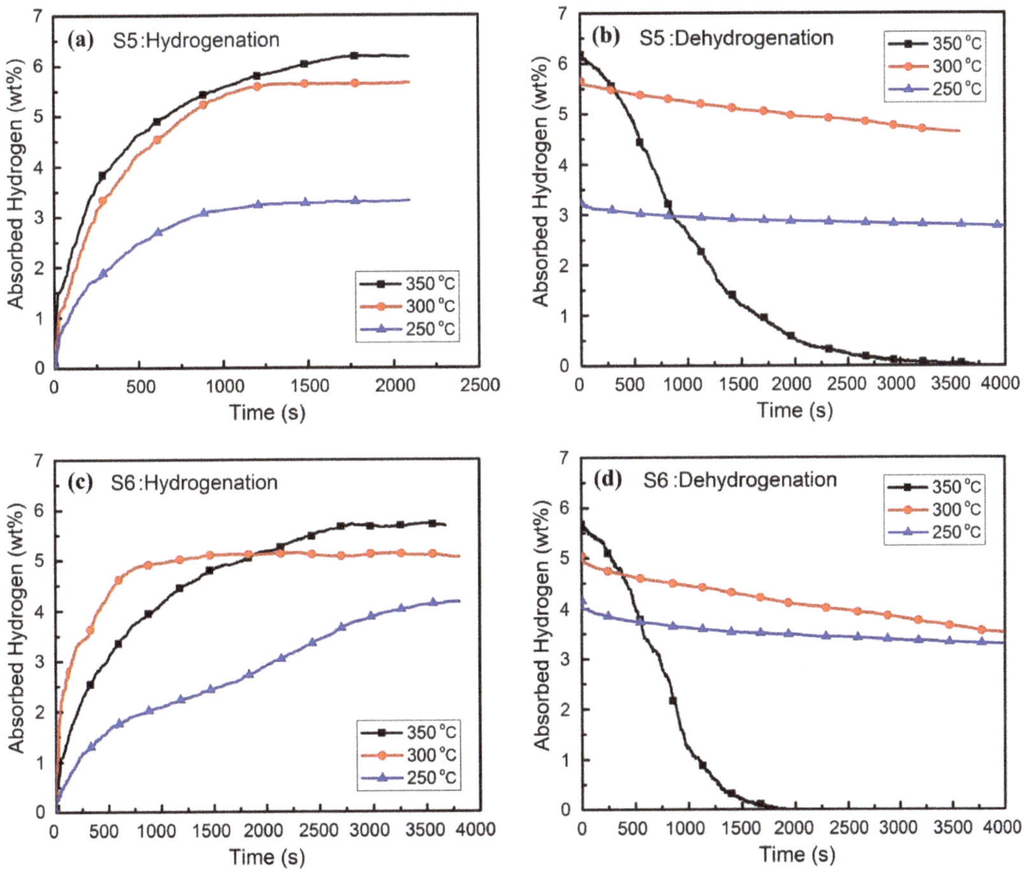

Figure 5. Hydrogenation kinetics at 30 bar and dehydrogenation kinetics at 10^{-1} mbar for different temperatures: (**a**,**b**) S5 sample (third pass of TPCAE at 250 °C); (**c**,**d**) S6 sample (fourth pass of TPCAE at 200 °C).

The hydrogenation kinetic for the S5 sample at 300 °C and at 350 °C is similar. At 250 °C, the maximum absorption after the third pass of TPCAE is about 3.3 wt.% which is lower than that 4.3 wt.%, which was achieved after the second pass of TPCAE (S3 sample), but the three passes deformed sample reaches its maximum capacity more quickly, which confirms the enhanced kinetic of the hydrogenation process at 250 °C. After the third pass of TPCAE (Figure 5b), full desorption of the sample is achieved at 350 °C with a longer time span in comparison to the sample after the second pass of TPCAE at 250° C (Figure 3b). By comparing the dehydrogenation kinetic of S5 sample (Figure 5b) with S4 sample (Figure 3d), it is obvious that the trend is almost identical at all temperatures.

As seen in Figure 5c, the fourth pass of the TPCAE (S6 sample) decreases the absorption capacity of the sample at 350 °C while it improves the hydrogenation kinetic at 300 °C. However, the maximum absorption capacity at 250 °C is improved after the fourth pass of TPCAE and reached to the best achievable one which was observed in S3 sample (compare Figure 5c with Figure 3a). Likewise, the dehydrogenation kinetic of S6 sample (Figure 5d) is close to that of S3 (Figure 3b).

It can be said that in all the processed samples, a significant improvement for hydrogenation behaviors, especially at 350 °C and 300 °C, and dehydrogenation behaviors at 350 °C is achieved, in comparison to the as cast sample. This is due to the increase of internal defects, especially grain boundaries, during TPCAE processing which act as the pathways for diffusion of hydrogen atoms and nucleation sites for hydride phase transformation. As mentioned in Table 3, the mean grain size is reduced from 550 μm for the as cast sample to the range of 2.19-13.54 μm for the TPCAE processed samples.

It is worth noting that twin boundaries cannot be as effective as grain boundaries for the diffusion of hydrogen atoms due to the lack of amorphous regions at the twin planes [40]. Therefore, the twinning mechanism is not beneficial as much as recrystallization for hydrogenation and dehydrogenation behaviors.

4. Discussion

To comprehensively evaluate the effect of TPCAE processing routes on the Hydrogenation of the AZ91 alloy, it is important to consider the kinetic of hydrogenation thoroughly, in respect of the nucleation and the growth of the transformed hydride phase. Generally, the formation of a hydride phase from a metal base is a complex process. Typically, it can be divided into four simplified reactions [41]: (i) dissociation/adsorption of the H_2 molecules; (ii) surface penetration of the H atoms; (iii) bulk diffusion of the H atoms; (iv) hydride formation. In this regard, Johnson–Mehl–Avrami (JMA) equation has been proven very versatile in explaining the hydrogenation kinetic of a variety of metal-hydrogen systems. Considering this approach, the transformed fraction of hydride phase (α) is given vs. time (t) as:

$$\alpha = 1 - exp\{-(kt)^n\} \quad (1)$$

where k is the kinetic rate constant (nucleation and growth rates) and n is referred to as the Avrami exponent describing the type of nucleation and the growth dimensionality [42]. By taking the natural logarithm of both sides of Equation (1), it becomes:

$$ln[-ln(1-\alpha)] = lnk + n\ lnt. \quad (2)$$

Considering $ln\ k$ as the intersection of a linear curve with the vertical axis, the kinetic rate constant can be obtained. The kinetic rate constant varies with temperature in the Arrhenius form, as [21]:

$$k = k_o exp(-Q/RT) \quad (3)$$

where Q is the apparent activation energy, T is the temperature, R is the gas constant, and k_o is a constant independent from temperature. By plotting the lnk versus $1/T$, the apparent activation energy is found from the slope of linear regression. The calculations for S4 are illustrated in Figure 6, as an example.

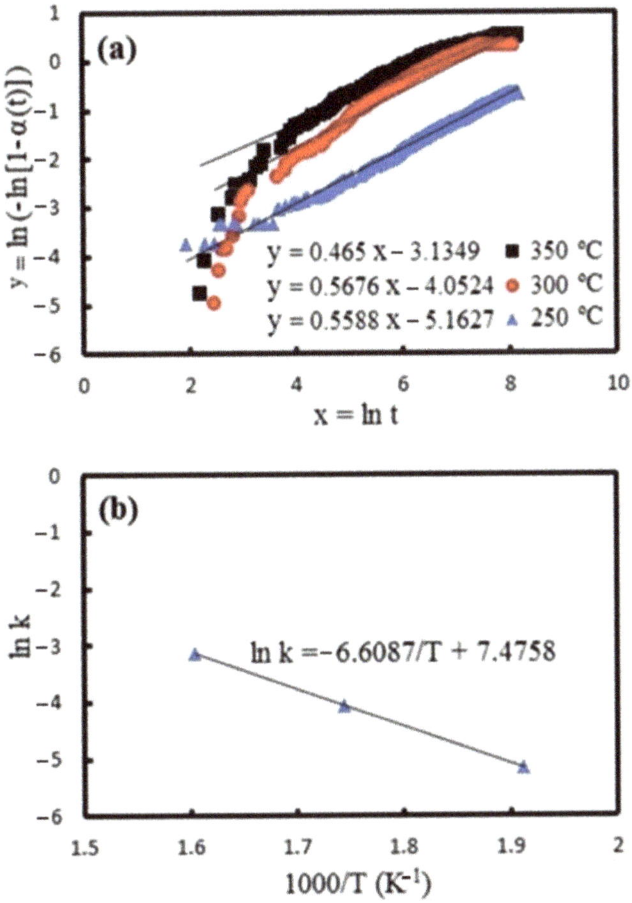

Figure 6. Determination of apparent activation energy (*Q*) for S4 sample (second pass of TPCAE at 200 °C): (**a**) plotting $ln[-ln(1-\alpha)]$ vs. lnt at different temperatures; (**b**) Arrhenius plot of the kinetic rate constants.

The kinetic parameters obtained by the JMA equation using the kinetic data are given in Tables 4 and 5 for all tested samples.

Table 4. Avrami exponent of the JMA model during hydrogenation of different samples at various temperatures.

Sample	n_{250}	n_{300}	n_{350}
As cast	0.73	0.48	0.39
S1	0.52	0.55	0.40
S2	0.72	0.4	0.66
S3	0.57	0.58	0.46
S4	0.56	0.57	0.47
S5	0.58	0.67	0.59
S6	0.6	0.28	0.59

As shown in Table 4, the values of n are around 0.5, denoting that the growth process is one dimensional while the rate limiting step is the diffusion of hydrogen [43]. The obtained values are consistent with that of reported by Andreasen [42], who calculated the kinetic parameters for the hydrogenation of ball-milled MgH_2-10wt.%$Mg_{17}Al_{12}$ composite. Since the n value is almost the same for all the processing conditions and there is not any considerable difference between the n values, it seems that the rate-limiting step does not change with increasing the TPCAE passes and/or processing temperature.

The calculated kinetic rate constants (Table 5), are quite consistent with those calculated by Jorge et al. [21] for an AZ91 Mg alloy after processing by ECAP under different conditions.

Table 5. Kinetic rate constants of the JMA model for the hydrogenation at different temperatures and apparent activation energy of the hydrogenation for different samples.

Sample	k_{250} [10^{-2} s^{-1}]	k_{300} [10^{-2} s^{-1}]	k_{350} [10^{-2} s^{-1}]	Apparent Activation Energy (kJ/mol)
As cast	0.08	0.69	1.7	82.9
S1	0.8	1.7	5.4	51.2
S2	0.2	4.3	1.2	48.2
S3	1.2	1.4	4.2	34.2
S4	0.6	1.7	4.3	54.9
S5	1.1	1.4	2.9	27
S6	0.6	15.9	1.6	29.8

By increasing the kinetic rate constant of the JMA model, a higher rate for nucleation and growth of transformed hydride phase is expected [21]. For the as cast sample, the kinetic rate constant decreases by one digit per each 50 °C of temperature reduction. This shows that the temperature reduction has an inverse effect on the hydrogenation kinetic for the cast alloy. For the TPCAE processed samples (with an exception of S6 sample), the kinetic rate constants for the hydrogenation at 350 °C and 300 °C, are in the same order of magnitude with a small difference, indicating that nucleation and growth rates at 300 °C are comparable to those at 350 °C, which conforms the success of TPCAE processing, to moderate the effect of lowering temperature on the nucleation and growth rates. The kinetic rate constant of hydrogenation at 300 °C for the S6 sample (fourth pass of TPCAE at 200 °C) is one order of magnitude higher than that at 350 °C, approving the effectiveness of the processing route of S6 for improvement in the hydrogenation kinetic at 300 °C. For the hydrogenation at 250 °C, S3 and S5 samples have the highest kinetic rate constants, which is attributed to the occurrence of recrystallization during the TPCAE processing that results in the grain sizes lower than 3 μm (see Table 3). Besides, for both S3 and S5 samples, the last pass of the TPCAE processing was performed at 250 °C (see Table 2), showing that the optimum TPCAE processing temperature to have an acceptable hydrogenation kinetic at low temperature is 250 °C.

In terms of the activation energies in Table 5, it is obvious that the highest amount corresponds to the cast sample. This is consistent with the obtained value of 81 kJ/mol by Andreasen [42], on the hydrogenation activation energy of ball-milled MgH_2-10 wt.%$Mg_{17}Al_{12}$ composite. The range of activation energy from 27 to 55 kJ/mol for the TPCAE processed samples shows a significant improvement in comparison to the range of 90-140 kJ/mol usually reported for the activated Mg powders [42,44], and 72 kJ/mol reported for the Mg thin films in the literature [45]. The values at the upper bound of the range correspond to the S1 and S4. The lowest activation energies are related to S5 (third pass) and S6 (fourth pass) that have the highest amount of thermomechanical processing. As mentioned, the internal defects such as the grain boundaries, the dislocations, and the precipitate interfaces facilitate the diffusion of the hydrogen atoms through the material bulk and provide nucleation sites for the formation of hydride phase, leading to the reduction of the activation energy. Therefore, the activation energy variations can be interpreted by the microstructure evolution during the TPCAE processing.

Considering all the aforementioned parameters, the highest activation energy is expected for the as cast sample, due to its largest grain size and its lowest dislocation density. After a first pass of TPCAE, grain size decreases significantly, a huge number of dislocations generate in the microstructure, and initial β-phase precipitates start to appear (Table 3), so a lower activation energy is observed.

By further processing the sample (the second pass of TPCAE), grain size decreased and the volume fraction of precipitates increased that result in a further decrease in the activation energy. However, the lowest activation energy for the second pass samples is achieved at the processing temperature of 250 °C (S3 sample) due to its lower processing temperature than the S2 sample (second pass of TPCAE at 300 °C) in which the grain growth is hindered by the lower diffusion rate of Mg atoms, and due to the more homogenous microstructure than the S4 sample (second pass of TPCAE at 200 °C), more detail of the microstructural investigations can be found in Section 3.1. By continuing the TPCAE processing to the third (S5 sample) and fourth pass (S6 sample) of the process, a finer microstructure, and a higher amount of β-phase precipitates are observed (see Table 3). A higher volume fraction of β-phase precipitates with fine dispersion, results in the higher contribution of interfaces in the diffusion of the hydrogen atoms into the material bulk [21], which would improve the diffusion of hydrogen atoms and reduction of the activation energy. Additionally, the β-phase precipitates, acting as a catalyst, reduce the H_2 dissociation energy [46], and may act as nucleation points for the hydride phase [39]. From all the above, it is expected to have a lower activation energy for the samples that have higher thermomechanical works and have their last pass of the TPCAE process at lower temperatures.

However, on the contrary to the S2 and S3 samples, the S4 sample, with the second pass of processing at 200 °C, has a relatively high activation energy as S1. The reason is that, although the processing at a low temperature of 200 °C may lead to a high amount of precipitation and dislocation density, such a sudden reduction of temperature causes a more bimodal and non-homogenous microstructure, as mentioned before, which is not effective for the reduction of activation energy. In fact, implementing a sufficient thermomechanical work while there is a step-by-step temperature reduction throughout the passes leads to form a finer microstructure with a homogenous distribution of the precipitates, playing the major role in reducing the activation energy and improving the hydrogenation kinetic. This explains the lowest activation energies for S5 sample (third pass of TPCAE at 250 °C) and S6 sample (fourth pass of TPCAE at 200 °C).

Generally, considering the absorption and desorption behavior of different samples with different conditions, it could be stated that the TPCAE processing has made a remarkable improvement on the kinetics of hydrogenation and dehydrogenation of cast AZ91 alloy, specifically when the gradual reduction of processing temperature is applied during the implementation of sufficient thermomechanical work level.

5. Conclusions

In comparison to the cast condition, significant improvements are obtained via TPCAE processing in terms of both increasing the storage capacity and kinetics of absorption/desorption. For hydrogenation, a significant improvement is achieved at 350 °C and 300 °C, but to a lower extent at 250 °C. The hydrogenation behavior at 300 °C becomes closer to the behavior at 350 °C. On the other hand, the improvement in dehydrogenation is made particularly at 350 °C. Applying sufficient level of thermomechanical work during the step-by-step temperature reduction results in the formation of a homogenous and fine-grained microstructure, including fine dispersion of β-$Mg_{17}Al_{12}$ precipitates. Therefore, due to the increase of the internal defects, which are appropriate for diffusion of hydrogen atoms, remarkable results are obtained, as below:

- The maximum storage capacity of 6.1 wt.% is obtained at a time span below 2000 s.
- The activation energy for hydrogenation is reduced from 82.9 kJ/mol to the range of 27–54.9 kJ/mol. The results show that the TPCAE processing is capable of reducing the activation energy to lower than 30 kJ/mol.

- The complete dehydrogenation is done at 350 °C in a time range of 1000 s up to maximum 2500 s.

Author Contributions: Investigation, M.A., R.E. and E.B.; writing—original draft preparation, M.A. and R.E.; formal analysis and data curation, M.A.; project administration, funding acquisition, and conceptualization, R.E.; supervision, R.E. and E.B.; writing—review and editing, E.B. All authors have read and agreed to the published version of the manuscript.

Funding: This research was funded by SHIRAZ UNIVERSITY, grant number 99-GR-ENG 15.

Conflicts of Interest: The authors declare no conflict of interest. The funders had no role in the design of the study; in the collection, analyses, or interpretation of data; in the writing of the manuscript; or in the decision to publish the results.

References

1. Varin, R.A.; Czujko, T.; Wronski, Z.S. *Nanomaterials for Solid State Hydrogen Storage*; Springer Science & Business Media: New York, NY, USA, 2009; pp. 3–5. [CrossRef]
2. Schlapbach, L.; Züttel, A. Hydrogen-storage materials for mobile applications. In *Materials for Sustainable Energy: A Collection of Peer-Reviewed Research and Review Articles from Nature Publishing Group*; World Scientific: Singapore, 2011; pp. 265–270. [CrossRef]
3. Yang, Y.; Xiong, X.; Chen, J.; Peng, X.; Chen, D.; Pan, F. Research advances in magnesium and magnesium alloys worldwide in 2020. *J. Magnes. Alloys* **2021**, *9*, 705–747. [CrossRef]
4. Huot, J.; Liang, G.; Boily, S.; Van Neste, A.; Schulz, R. Structural study and hydrogen sorption kinetics of ball-milled magnesium hydride. *J. Alloys Compd.* **1999**, *293*, 495–500. [CrossRef]
5. Song, M.; Lee, J.Y. A study of the hydriding kinetics of Mg-(10–20 w/o) LaNi5. *Int. J. Hydrogen Energy* **1983**, *8*, 363–367. [CrossRef]
6. Soyama, J.; Triques, M.R.M.; Leiva, D.R.; Junior, A.M.J.; da Silva, E.P.; Pinto, H.C.; Bolfarini, C.; Kiminami, C.S.; Botta, W.J. Hydrogen storage in heavily deformed ZK60 alloy modified with 2.5 wt.% Mm addition. *Int. J. Hydrogen Energy* **2016**, *41*, 4177–4184. [CrossRef]
7. Huang, S.-J.; Chiu, C.; Chou, T.-Y.; Rabkin, E. Effect of equal channel angular pressing (ECAP) on hydrogen storage properties of commercial magnesium alloy AZ61. *Int. J. Hydrogen Energy* **2018**, *43*, 4371–4380. [CrossRef]
8. Li, X.; Lin, Y. Influence of Al doping on stability of Mg1− xTix and their hydrides. *Acta Phys. Sin.* **2013**, *62*, 138801. [CrossRef]
9. Au, M. Hydrogen storage properties of magnesium based nanostructured composite materials. *Mater. Sci. Eng. B* **2005**, *117*, 37–44. [CrossRef]
10. Leiva, D.R.; Fruchart, D.; Bacia, M.; Girard, G.; Skryabina, N.; Villela, A.C.; Miraglia, S.; Santos, D.S.; Botta, W.J. Mg alloy for hydrogen storage processed by SPD. *Int. J. Mater. Res.* **2009**, *100*, 1739–1746. [CrossRef]
11. Zaluski, L.; Zaluska, A.; Tessier, P.; Strom-Olsen, J.; Schulz, R. Nanocrystalline hydrogen absorbing alloys. In Proceedings of the Materials Science Forum, 10 July 1996; p. 2. [CrossRef]
12. Révész, Á.; Gajdics, M.; Schafler, E.; Calizzi, M.; Pasquini, L. Dehydrogenation-hydrogenation characteristics of nanocrystalline Mg_2Ni powders compacted by high-pressure torsion. *J. Alloys Compd.* **2017**, *702*, 84–91. [CrossRef]
13. Skripnyuk, V.; Rabkin, E.; Bendersky, L.; Magrez, A.; Carreño-Morelli, E.; Estrin, Y.J. Hydrogen storage properties of as-synthesized and severely deformed magnesium–multiwall carbon nanotubes composite. *Int. J. Hydrogen Energy* **2010**, *35*, 5471–5478. [CrossRef]
14. Khani, S.; Salehi, M.; Samim, H.; Aboutalebi, M.; Palkowski, H. The effect of severe plastic deformation on the microstructure and mechanical properties of as-cast AZ31. *Iran. J. Mater. Sci. Eng.* **2016**, *13*, 29–38.
15. Malik, A.; Wang, Y.; Nazeer, F.; Khan, M.A.; Ali, T.; Ain, Q.T. Effect of pre-straining on twinning, texture and mechanical behavior of magnesium alloys A-review. *J. Mater. Res. Technol.* **2020**, *9*, 14478–14499. [CrossRef]
16. Skripnyuk, V.; Rabkin, E.; Estrin, Y.; Lapovok, R. Improving hydrogen storage properties of magnesium based alloys by equal channel angular pressing. *Int. J. Hydrogen Energy* **2009**, *34*, 6320–6324. [CrossRef]
17. Chiu, C.; Huang, S.-J.; Chou, T.-Y.; Rabkin, E. Improving hydrogen storage performance of AZ31 Mg alloy by equal channel angular pressing and additives. *J. Alloys Compd.* **2018**, *743*, 437–447. [CrossRef]
18. Jorge, A.M., Jr.; Prokofiev, E.; de Lima, G.F.; Rauch, E.; Veron, M.; Botta, W.J.; Kawasaki, M.; Langdon, T.G. An investigation of hydrogen storage in a magnesium-based alloy processed by equal-channel angular pressing. *Int. J. Hydrogen Energy* **2013**, *38*, 8306–8312. [CrossRef]
19. Edalati, K.; Yamamoto, A.; Horita, Z.; Ishihara, T. High-pressure torsion of pure magnesium: Evolution of mechanical properties, microstructures and hydrogen storage capacity with equivalent strain. *Scr. Mater.* **2011**, *64*, 880–883. [CrossRef]
20. Hongo, T.; Edalati, K.; Arita, M.; Matsuda, J.; Akiba, E.; Horita, Z. Significance of grain boundaries and stacking faults on hydrogen storage properties of Mg_2Ni intermetallics processed by high-pressure torsion. *Acta Mater.* **2015**, *92*, 46–54. [CrossRef]
21. Jorge, A.M., Jr.; Prokofiev, E.; Triques, M.R.M.; Roche, V.; Botta, W.J.; Kiminami, C.S.; Raab, G.I.; Valiev, R.Z.; Langdon, T.G. Effect of cold rolling on the structure and hydrogen properties of AZ91 and AM60D magnesium alloys processed by ECAP. *Int. J. Hydrogen Energy* **2017**, *42*, 21822–21831. [CrossRef]

22. Skripnyuk, V.; Buchman, E.; Rabkin, E.; Estrin, Y.; Popov, M.; Jorgensen, S. The effect of equal channel angular pressing on hydrogen storage properties of a eutectic Mg–Ni alloy. *J. Alloys Compd.* **2007**, *436*, 99–106. [CrossRef]
23. Skripnyuk, V.; Rabkin, E.; Estrin, Y.; Lapovok, R. The effect of ball milling and equal channel angular pressing on the hydrogen absorption/desorption properties of Mg–4.95 wt.% Zn–0.71 wt.% Zr (ZK60) alloy. *Acta Mater.* **2004**, *52*, 405–414. [CrossRef]
24. Asselli, A.; Leiva, D.; Huot, J.; Kawasaki, M.; Langdon, T.; Botta, W. Effects of equal-channel angular pressing and accumulative roll-bonding on hydrogen storage properties of a commercial ZK60 magnesium alloy. *Int. J. Hydrogen Energy* **2015**, *40*, 16971–16976. [CrossRef]
25. Botta, W.; Jorge, A., Jr.; Veron, M.; Rauch, E.; Ferrie, E.; Yavari, A.; Huot, J.; Leiva, D. H-sorption properties and structural evolution of Mg processed by severe plastic deformation. *J. Alloys Compd.* **2013**, *580*, S187–S191. [CrossRef]
26. Jorge, A.M., Jr.; de Lima, G.F.; Triques, M.R.M.; Botta, W.J.; Kiminami, C.S.; Nogueira, R.P.; Yavari, A.R.; Langdon, T.G. Correlation between hydrogen storage properties and textures induced in magnesium through ECAP and cold rolling. *Int. J. Hydrogen Energy* **2014**, *39*, 3810–3821. [CrossRef]
27. Huot, J.; Skryabina, N.Y.; Fruchart, D. Application of severe plastic deformation techniques to magnesium for enhanced hydrogen sorption properties. *Metals* **2012**, *2*, 329–343. [CrossRef]
28. Abdi, M.; Ebrahimi, R. Twin parallel channel angular extrusion as a development of ECAE in parallel channels. *SN Appl. Sci.* **2020**, *2*, 1–10. [CrossRef]
29. Schneider, C.A.; Rasband, W.S.; Eliceiri, K.W. NIH Image to ImageJ: 25 years of image analysis. *Nat. Methods* **2012**, *9*, 671–675. [CrossRef] [PubMed]
30. Abdi, M.; Ebrahimi, R. Microstructure Evolution of AZ91 Alloy Processed by Twin Parallel Channel Angular Extrusion Technique. *J. Mater. Eng. Perform.* **2022**, *31*, 5358–5373. [CrossRef]
31. Asselli, A.A.C.; Hébert, N.B.; Huot, J. The role of morphology and severe plastic deformation on the hydrogen storage properties of magnesium. *Int. J. Hydrogen Energy* **2014**, *39*, 12778–12783. [CrossRef]
32. Dini, H.; Svoboda, A.; Andersson, N.-E.; Ghassemali, E.; Lindgren, L.-E.; Jarfors, A.E. Optimization and validation of a dislocation density based constitutive model for as-cast Mg-9% Al-1% Zn. *Mater. Sci. Eng. A* **2018**, *710*, 17–26. [CrossRef]
33. Khani, S.; Aboutalebi, M.; Salehi, M.; Samim, H.; Palkowski, H. Microstructural development during equal channel angular pressing of as-cast AZ91 alloy. *Mater. Sci. Eng. A* **2016**, *678*, 44–56. [CrossRef]
34. Wang, Q.; Chen, Y.; Liu, M.; Lin, J.; Roven, H.J. Microstructure evolution of AZ series magnesium alloys during cyclic extrusion compression. *Mater. Sci. Eng. A* **2010**, *527*, 2265–2273. [CrossRef]
35. Yuan, Y.; Guo, Q.; Sun, J.; Liu, H.; Xu, Q.; Wu, Y.; Song, D.; Jiang, J.; Ma, A. High mechanical properties of AZ91 Mg alloy processed by equal channel angular pressing and rolling. *Metals* **2019**, *9*, 386. [CrossRef]
36. Peng, W.; Lan, Z.; Wei, W.; Xu, L.; Guo, J. Investigation on preparation and hydrogen storage performance of $Mg_{17}Al_{12}$ alloy. *Int. J. Hydrogen Energy* **2016**, *41*, 1759–1765. [CrossRef]
37. Jia, Y.; Sun, C.; Shen, S.; Zou, J.; Mao, S.S.; Yao, X. Combination of nanosizing and interfacial effect: Future perspective for designing Mg-based nanomaterials for hydrogen storage. *Renew. Sustain. Energy Rev.* **2015**, *44*, 289–303. [CrossRef]
38. Floriano, R.; Leiva, D.; Melo, G.; Ishikawa, T.; Huot, J.; Kaufman, M.; Figueroa, S.; Mendoza-Zélis, L.A.; Damonte, L.C.; Botta, W. Low temperature rolling of AZ91 alloy for hydrogen storage. *Int. J. Hydrogen Energy* **2017**, *42*, 29394–29405. [CrossRef]
39. Amira, S.; Huot, J. Effect of cold rolling on hydrogen sorption properties of die-cast and as-cast magnesium alloys. *J. Alloys Compd.* **2012**, *520*, 287–294. [CrossRef]
40. Hui, J.; Zhang, X.; Liu, T.; Liu, W.; Wang, B. First-principles calculation of twin boundary energy and strength/embrittlement in hexagonal close-packed titanium. *Mater. Des.* **2022**, *213*, 110331. [CrossRef]
41. Töpler, J.; Buchner, H.; Säufferer, H.; Knorr, K.; Prandl, W. Measurements of the diffusion of hydrogen atoms in magnesium and Mg_2Ni by neutron scattering. *J. Less Common Met.* **1982**, *88*, 397–404. [CrossRef]
42. Andreasen, A. Hydrogenation properties of Mg–Al alloys. *Int. J. Hydrogen Energy* **2008**, *33*, 7489–7497. [CrossRef]
43. Karty, A.; Grunzweig-Genossar, J.; Rudman, P. Hydriding and dehydriding kinetics of Mg in a Mg/Mg_2Cu eutectic alloy: Pressure sweep method. *J. Appl. Phys.* **1979**, *50*, 7200–7209. [CrossRef]
44. Galey, B.; Auroux, A.; Sabo-Etienne, S.; Dhaher, S.; Grellier, M.; Postole, G. Improved hydrogen storage properties of Mg/MgH_2 thanks to the addition of nickel hydride complex precursors. *Int. J. Hydrogen Energy* **2019**, *44*, 28848–28862. [CrossRef]
45. Johansson, M.; Ostenfeld, C.W.; Chorkendorff, I. Adsorption of hydrogen on clean and modified magnesium films. *Phys. Rev. B* **2006**, *74*, 193408. [CrossRef]
46. Crivello, J.-C.; Nobuki, T.; Kato, S.; Abe, M.; Kuji, T. Hydrogen absorption properties of the γ-$Mg_{17}Al_{12}$ phase and its Al-richer domain. *J. Alloys Compd.* **2007**, *446*, 157–161. [CrossRef]

crystals

Article

Experimental Study on Laser Lap Welding of Aluminum–Steel with Pre-Fabricated Copper–Nickel Binary Coating

Hua Zhang *, Huiyan Gu and Dong Ma

School of Mechanical Engineering, Nantong University, Nantong 226019, China; ghy9909@163.com (H.G.); madong0201@126.com (D.M.)
* Correspondence: ntuzhh@ntu.edu.cn; Tel.:+86-0513-85012674

Abstract: In order to solve the problem of poor weld quality caused by brittle metal compounds in the welding of dissimilar metals between aluminum and steel, a pre-welding treatment method of prefabricated copper–nickel binary coating between aluminum and steel has been proposed. Laser lap welding tests and weld performance tests were conducted using 6061 aluminum alloy and DP590 duplex steel with a thickness of 0.5 mm as base materials, with steel on top and aluminum on bottom. The research results indicate that the prefabricated copper–nickel binary coating can effectively suppress the formation of brittle phase compounds of Fe and Al; the increase of copper and nickel elements is beneficial for the formation of tough compounds such as $(Fe, Cu, Ni)_3Al$, $(Fe, Cu, Ni)Al_3$, and $CuAl_5$ in the weld zone; when the thickness of the copper coating is 155 μm and the thickness of the nickel coating is 110 μm, the mechanical properties of the aluminum steel lap welding seam are the best, and the maximum shear force that can be withstood is 208.09 N, which is 56% higher than uncoated sample.

Keywords: laser welding; dissimilar metals; copper–nickel binary coating; mechanical properties; electroplating

1. Introduction

Automotive lightweighting, driven by the need for energy conservation and emission reduction, has become a critical focus for sustainable development in the automotive industry. However, single metal materials can no longer meet the modern automobile's multifaceted requirements, such as high strength, low weight, and corrosion resistance. Consequently, the development of dissimilar metal composite structures has emerged as a key direction for realizing automotive lightweighting. By combining the advantages of different metal materials, dissimilar metal composites can significantly enhance structural strength, reduce weight, and improve durability, thus providing a new technological pathway for the sustainable development of the automotive industry [1–5].

The reliable welding of aluminum–steel dissimilar metals is one of the key factors in industrial lightweighting. However, due to the significant differences in the physical properties of aluminum and steel, welding joints often suffer from defects such as cracks and porosity. Additionally, brittle intermetallic compounds are abundant in the weld zones, reducing the quality of aluminum–steel welded joints [6–11]. Currently, the addition of alloying elements between aluminum and steel can improve the wettability of the fusion zone, alter the structure and composition of the weld zone, and suppress the formation of brittle intermetallic compounds [12–14].

Common methods for adding alloying elements between aluminum and steel include metal powders, alloy welding wires, and alloy foils. Li B et al. performed laser lap welding of 5083 aluminum alloy with DP780 steel using FeCoCrNiMn and Al-10Si-0.3Mg powders, respectively [15]. By comparing the macroscopic morphology, microstructure, hardness, and shear performance of the two types of welded joints, it was found that FeCoCrNiMn powder demonstrated significant advantages over Al-10Si-0.3Mg powder in reducing weld hardness and improving the shear performance of lap-welded joints. Esmaily H et al. conducted laser lap welding of reinforced 5456 aluminum alloy and stainless steel [16], discovering that adding BNi-2 powder transformed the weld microstructure from Fe-Al brittle intermetallic compounds to ternary Fe-Ni-Al brittle intermetallic compounds. Consequently, the weld joint achieved a shear strength of 114 MPa.

Yu G et al. performed a comparative study of laser welding-brazing for butt joints of aluminum alloys and galvanized steel [17]. They used five different filler materials: Al, $AlSi_5$, $AlMg_5$, $AlCu_6$, and $ZnAl_2$, analyzing their impact on the microstructure, mechanical properties, and weld quality of the joints. The results showed that selecting appropriate filler materials significantly improved the bonding strength, corrosion resistance, and toughness of the joints, especially at the aluminum alloy and galvanized steel interface. Yiyang H et al. studied the impact of Si content in brazing alloys on the morphology, microstructure, and mechanical properties of aluminum–steel welds using ER1050, AlSi5, AlSi10, and AlSi15 cored wires for laser welding of 6061 aluminum alloy and DP590 dual-phase steel [18]. They found that increasing Si content significantly enhanced the tensile performance of the welded joints, with AlSi10 wire achieving the best mechanical performance.

Yaochen S et al. employed copper and nickel foils for laser welding of steel/aluminum [19]. Their results indicated that the addition of copper and nickel improved the alloying reactions at the aluminum–steel interface, reduced the formation of brittle intermetallic compounds, and significantly enhanced the tensile strength, hardness, and toughness of the joints. Zhang Y et al. examined the effects of copper and vanadium foils on the brittleness mechanisms of aluminum/steel laser-welded joints [20], finding that these interlayers effectively improved the microstructure of the joints, mitigated the formation of brittle intermetallic compounds, and enhanced the mechanical properties. Li T et al. investigated the effects of laser coupled with titanium foil on the microstructure and mechanical properties of steel/aluminum fusion-welded joints [21]. Their results demonstrated that laser welding with titanium foil significantly improved the microstructure, forming a more uniform intermetallic compound layer, thereby enhancing the joint's strength and toughness.

The above methods, including the addition of alloy powders, welding wires, and alloy foils, have achieved notable success in improving the quality of aluminum–steel welds. However, the quantification and control of added alloying elements remain insufficient.

This study introduces the use of electrochemical deposition to pre-deposit copper and nickel coatings on aluminum and steel substrates. This approach, rarely explored in prior research, aims to achieve a uniform binary interlayer for aluminum–steel laser lap welding. The resulting binary metal coating, formed by overlapping the two materials, was used as an interlayer. Aluminum–steel laser lap welding was then performed. Tensile testing and microscopic observation of the weld structure were conducted to reveal the influence of the copper–nickel binary coating on the quality of aluminum–steel welds.

2. Materials and Methods

2.1. Experimental Equipments

The power supply used in the electrochemical deposition equipment for this experiment is the GKPT numerically controlled flat wave/bidirectional pulse adjustable power supply produced by Shenzhen Shicheng Electronic Technology Co., Ltd. (Shenzhen, China). This power supply enables the switching between DC and AC currents and features automatic timing of electrochemical deposition. It provides precise control over the electrochemical deposition process. The adjustable pulse frequency ranges from 1 to 5000 Hz, and the pulse duty cycle can be adjusted from 0% to 100% in the positive direction and from 0% to 30% in the negative direction.

The laser welding equipment used in this experiment is a multifunctional laser processing machine, model JHM-1GXY-500, produced by Wuhan Chutian Industrial Laser Equipment Co., Ltd. (Wuhan, China). The system utilizes an Nd:YAG laser, and the CNC workbench is controlled by a PA system for X and Y two-dimensional linkage, with a stroke of 200 mm × 150 mm. It is also equipped with a CCD coaxial monitoring system.

2.2. Experimental Materials

The materials used in the experiment were 6061 aluminum alloy and domestically produced cold-rolled DP590 dual-phase steel, both with dimensions of 100 mm × 100 mm × 0.5 mm. The chemical composition of DP590 dual-phase steel is shown in Table 1, and the chemical composition of 6061 aluminum alloy is presented in Table 2.

Table 1. Chemical composition of DP 590 (mass score/%).

Material	C	Mn	P	Si	Ni	S	Cr	Fe
DP590	0.150	2.500	0.040	0.600	-	0.015	0.020	Bal.

Table 2. Chemical composition of 6061 aluminum alloy (mass score/%).

Material	Cu	Mn	Mg	Zn	Cr	Ti	Si	Fe	Al
6061	0.15	0.15	0.8	0.25	0.04	0.15	0.4	0.7	Bal.

2.3. Pre-Deposited Copper–Nickel Binary Coating

The DP590 dual-phase steel and 6061 aluminum alloy specimens were separately immersed in a nickel chloride electroplating solution and a copper sulphate electroplating solution to apply nickel and copper coatings on the welding zone surfaces. The current density was set to 6 A/dm^2, and the electroplating durations were 1 h, 2 h, and 3 h. The electroplated samples are shown in Figures 1 and 2.

Figure 1. Copper-plated 6061 aluminum alloy specimen.

Figure 2. Nickel-plated DP590 dual-phase steel specimen.

By varying the electrochemical deposition time to 1 h, 2 h, and 3 h, the resulting thicknesses of the copper and nickel coatings were measured, as shown in Table 3.

Table 3. Thickness variation of copper–nickel binary plating at different times.

Serial Number	Time (h)	Cu Coating Thickness (μm)	Ni Coating Thickness (μm)
1	1	60	40
2	2	120	75
3	3	155	110

2.4. *Steel/Aluminum Laser Lap Welding Tests*

The laser lap welding experiment was conducted using a lap joint configuration with steel on top and aluminum underneath. Welding was carried out on the pre-coated aluminum–steel dissimilar materials with side-blown shielding gas. The airflow direction formed a 30° angle with the laser beam, as illustrated in Figure 3.

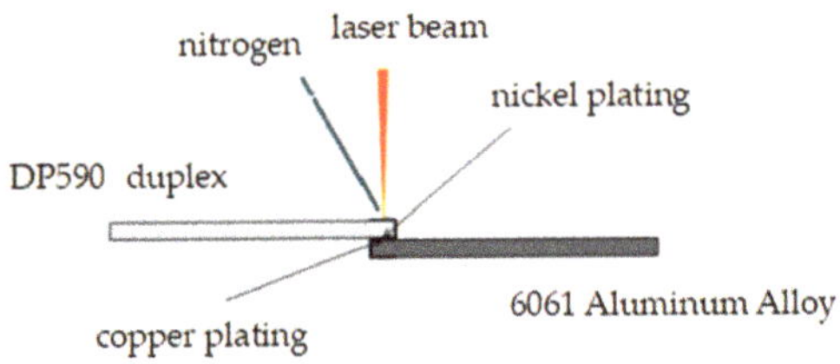

Figure 3. Schematic diagram of laser welding.

To investigate the effect of the copper–nickel binary coating thickness on the quality of aluminum–steel welds, experiments were conducted with coating thickness as the control variable. A Nd:YAG laser was used as the heat source, with the laser beam positioned perpendicular to the workpiece. The welding speed was set at 100 mm/min, the defocus amount was zero, the single pulse energy was 9 J, the pulse width was fixed at 6 ms, and the frequency was 15 Hz. Additionally, the welding process employed a pulse linear scanning strategy, where the laser beam scanned in a straight-line path. The spot overlap ratio was set to 80%, ensuring stable fusion and reducing defects.

2.5. *Quality Assessment of Aluminum–Steel Lap Welds*

2.5.1. Mechanical Performance Testing

The aluminum–steel lap-welded specimens were fabricated into corresponding tensile test samples in accordance with ASTM E8/E8M-21. Each welded plate was processed using wire-cut electrical discharge machining (EDM) to prepare three sets of tensile test specimens and two metallographic specimens, ensuring statistical reliability. The specific parameters of the tensile test specimens and the wire-cut EDM processing method are shown in Figure 4. Static lap shear tests were conducted at room temperature with a testing speed of 5 mm/min. An electronic universal tensile testing machine (CMT5105, MTS, Orlando, FL, USA) was used to measure the tensile strength of the specimens.

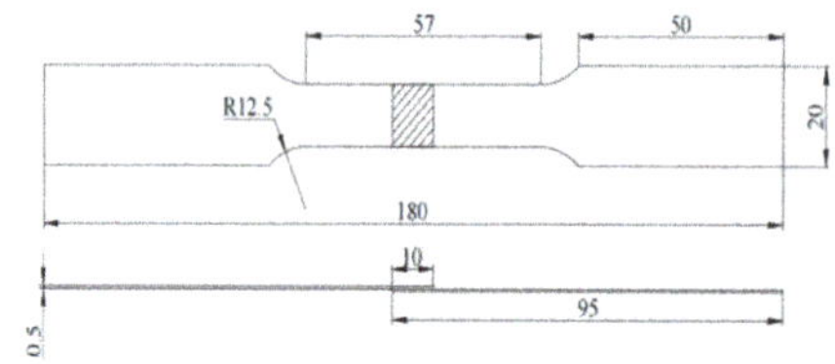

(a) Dimensional drawing of tensile specimen (unit: cm).

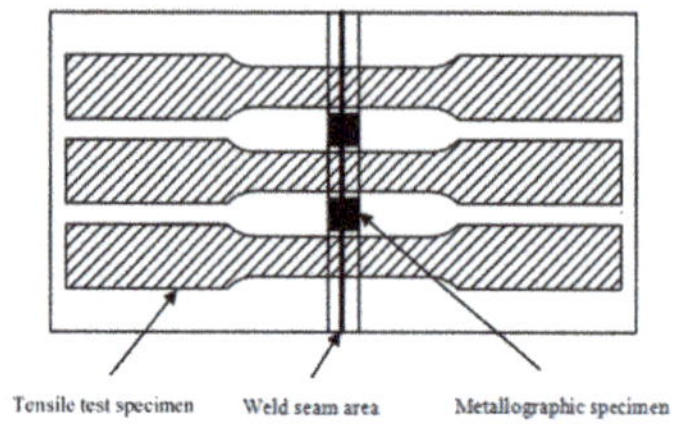

(b) Wire-cut EDM method

Figure 4. Dimensional drawing of tensile specimen and wire-cut EDM method.

2.5.2. Microstructural Observation

The welded specimens were wire-cut into 10 × 10 × 1 mm samples, followed by cleaning, hot mounting, grinding, and polishing. Metallographic specimens were prepared by etching with Keller's reagent (95 mL H_2O + 2.5 mL HNO_3 + 1.5 mL HCl + 1.0 mL HF).

Microstructure observations of the samples were carried out on the samples using a confocal microscope (Axio Scope A1, ZEISS, Oberkochen, Germany) and a scanning electron microscope (S-3400N, Hitachi, Tokyo, Japan). Energy-dispersive X-ray spectroscopy (S-3400N, Hitachi, Japan) was employed to perform line scan analysis of Cu and Ni element distribution across the top, middle, and bottom sections of the weld cross-section. To further investigate the primary composition of compounds in the weld, point scan analysis was conducted on six characteristic regions at the bottom of the weld pool to determine the elemental ratios of Fe, Al, Cu, and Ni.

3. Results and Analysis

3.1. Mechanical Performance Analysis of Welds

Figure 5 compares the tensile fracture diagrams of aluminum–steel lap welds with no coating and those with different thicknesses of the copper–nickel binary coating. As shown in Figure 5, the fracture locations of all four groups were within the aluminum–steel weld zone, with no damage observed in the base metal. This indicates that the mechanical performance of the welds is lower than that of the aluminum base metal.

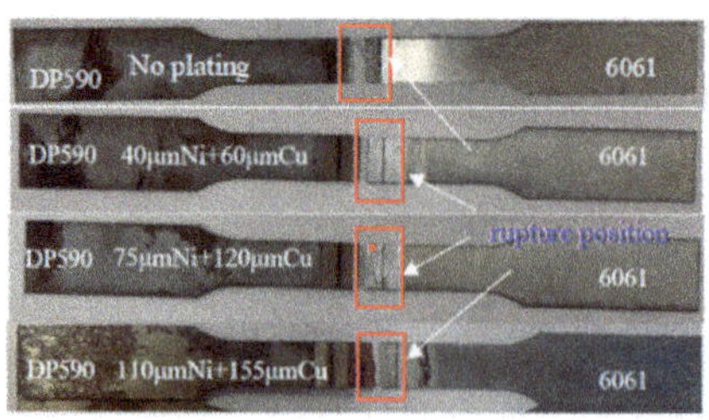

Figure 5. Tensile fracture of joints without plating and with different thicknesses of coating.

Figure 6 presents the load–displacement curves of aluminum–steel lap weld joints with and without coatings. The displacement in Figure 6 refers to the relative movement between the welded materials under applied shear force, typically representing the amount of deformation in the shear direction. The results demonstrate that the shear strength of the welds with coatings was superior to that of the welds without coatings. This improvement is primarily due to the addition of the copper–nickel coating, which inhibited the formation of brittle Fe-Al intermetallic compounds, thus enhancing the mechanical performance of the coated welds. When the thicknesses of the Cu and Ni coatings were 110 μm and 155 μm, respectively, the welds exhibited the highest shear strength, with a maximum shear force of 208 N, corresponding to a shear stress of 205 MPa. In contrast, for welds without coatings, the maximum shear force was 133 N, with a corresponding shear stress of 149 MPa.

The change around 80 N represents the end of the elastic deformation phase and the initiation of fracture. Up to this point, the welds with and without coatings exhibited similar elastic behaviour. However, in the plastic deformation phase, welds with coatings showed a higher ultimate shear strength but lower displacement before fracture, suggesting a trade-off between strength and ductility.

This enhancement is attributed to the increased copper and nickel content in the laser welding molten pool, where Cu and Ni atoms reacted metallurgically with Fe and Al atoms to form ductile compounds containing Cu and Ni. However, excessive coating thickness may promote the formation of additional intermetallics, leading to localized strain accumulation, promoting stress concentration and reducing the overall ductility of the welds. The microstructural heterogeneity introduced by thick coatings could further limit plastic deformation, resulting in early fracture initiation.

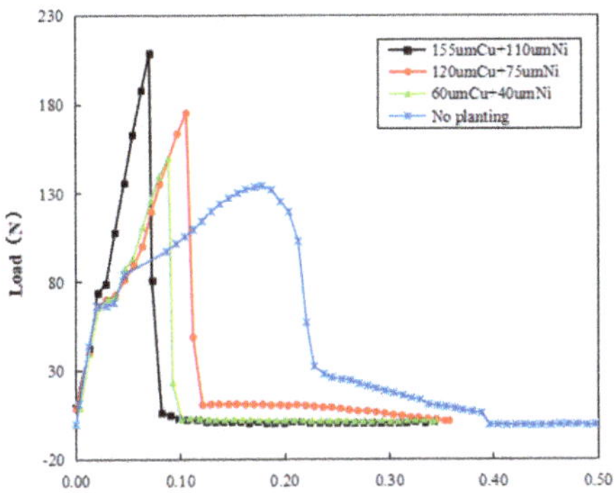

Figure 6. Tensile-shear load–displacement curves of joints without plating and with different thicknesses of plating.

Furthermore, a comparison with existing literature on aluminum–steel welding indicates that the shear stress values obtained in this study align with previously reported results. Guo et al. investigated aluminum–steel welded joints using laser welding-brazing with different filler-wire compositions, reporting shear force values ranging from 186 N to 203 N [22]. These values are consistent with the results obtained in this study, where the coated welds exhibited a shear force of 208 N. This comparison confirms the reliability of our experimental results and further highlights the effectiveness of Cu-Ni coatings in enhancing the shear strength of aluminum–steel welds.

3.2. Morphology Analysis of Welds

Figure 7 shows the cross-sectional morphology of aluminum–steel lap welds with varying thicknesses of copper–nickel binary coatings. When the Cu and Ni coating thicknesses were 60 μm and 40 μm, respectively, the thin intermediate metallic layer allowed residual heat from the melted DP590 dual-phase steel to penetrate the aluminum alloy below, resulting in a weld depth of 0.779 mm and a weld width of 1.116 mm. Significant ablation of the upper weld metal was observed. As the Cu and Ni coating thicknesses increased to 120 μm and 75 μm, respectively, a greater amount of Cu and Ni metal in the weld reduced heat transmission to the lower aluminum layer, leading to a flatter aluminum–steel fusion interface. The weld depth and width were measured at 0.821 mm and 0.947 mm, respectively, with no cracks observed. When the Cu and Ni coating thicknesses were further increased to 155 μm and 110 μm, minimal ablation of the upper weld metal occurred, allowing more heat transfer to the aluminum below. The weld depth reached 0.969 mm, and the weld width was 1.048 mm.

Figure 8 illustrates the weld morphology of aluminum–steel lap joints with varying copper–nickel coating thicknesses. At coating thicknesses of 60 μm for Cu and 40 μm for Ni, the weld fusion line appeared coarse with some stratification in the weld pool. As the coating thickness increased to 120 μm for Cu and 75 μm for Ni, the weld interface morphology improved, becoming tighter and smoother, although the fusion line contour remained irregular. At coating thicknesses of 155 μm for Cu and 110 μm for Ni, the weld interface morphology was optimal, featuring a smooth and compact fusion line, with no cracks observed in the weld pool.

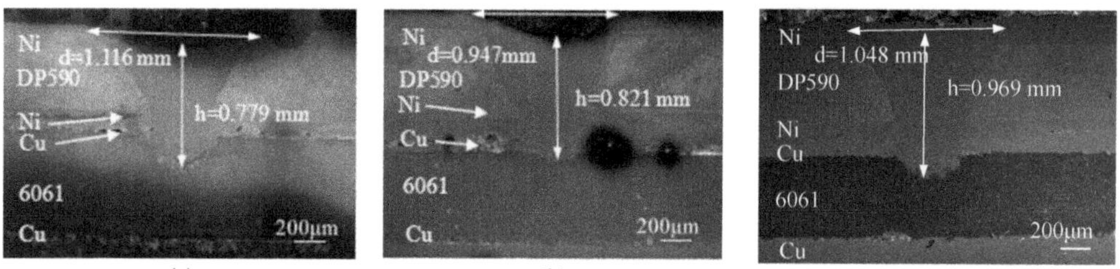

(a) (b) (c)

Figure 7. Cross-sectional morphology of welds in aluminum–steel lap welds with different thicknesses of preformed Cu-Ni binary coatings: (**a**) 60 μm Cu + 40 μm Ni; (**b**) 120 μm Cu + 75 μm Ni; (**c**) 155 μm Cu + 110 μm Ni.

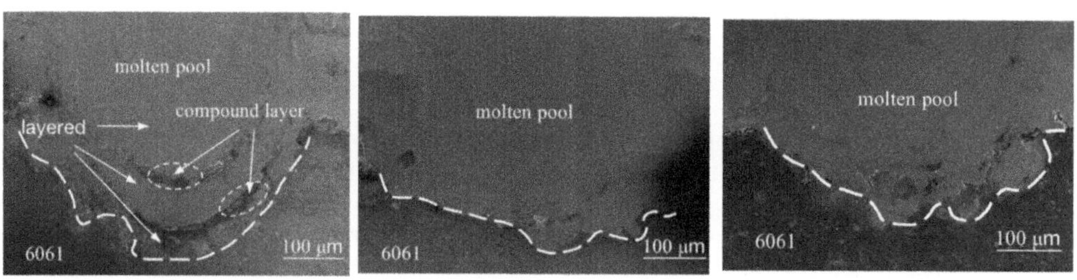

(a) (b) (c)

Figure 8. Weld morphology of aluminum–steel lap welds with different thicknesses of copper–nickel binary plating preformed: (**a**) 60 μm Cu + 40 μm Ni; (**b**) 120 μm Cu + 75 μm Ni; (**c**) 155 μm Cu + 110 μm Ni.

Comparisons revealed that as the coating thickness increased, the weld pool deepened, the morphology of the weld interface improved, stratification in the weld zone disappeared, and the fusion line became smoother and more regular, ultimately forming a "C" shape.

3.3. EDS Line Scan Analysis of Cu and Ni Elements in the Weld

Line scans were performed on the top, centre, and bottom sections of welds with 60 μm Cu + 40 μm Ni and 155 μm Cu + 110 μm Ni coatings. Figure 9 shows the line scan locations.

Figure 10 illustrates the distribution of Cu along the scanning path (X-axis: scan path length in micrometres; Y-axis: intensity of characteristic X-rays in counts per second). The results indicate that Cu was evenly distributed at the top and concentrated near the fusion line in the centre and bottom sections. This suggests that Cu serves as an interlayer, preventing the diffusion of Al atoms toward the steel side and suppressing the formation of brittle Fe-Al intermetallic compounds.

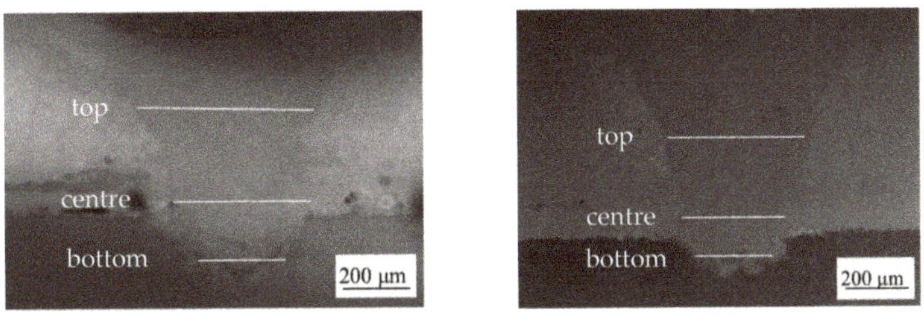

(**a**) (**b**)

Figure 9. Weld line scanning position of the welded joint with different thicknesses of copper–nickel binary plating preformed: (**a**) 60 μm Cu + 40 μm Ni; (**b**) 155 μm Cu + 110 μm Ni.

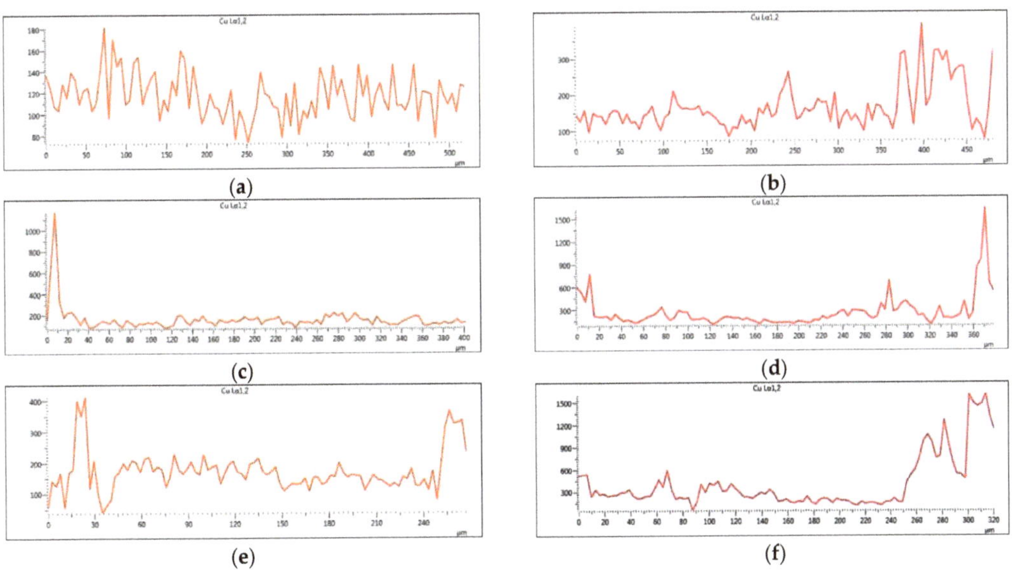

Figure 10. Cu element distribution in different section of welds with different copper–nickel binary: (**a**) top section of 60 μm Cu + 40 μm Ni weld; (**b**) top section of 155 μm Cu + 110 μm Ni weld; (**c**) centre section of 60 μm Cu + 40 μm Ni weld; (**d**) centre section of 155 μm Cu + 110 μm Ni weld; (**e**) bottom section of 60 μm Cu + 40 μm Ni weld; (**f**) bottom section of 155 μm Cu + 110 μm Ni weld.

Figure 11 presents the distribution of Ni with the same axis definitions as Figure 10. Compared to Cu, Ni exhibited more uniform diffusion, with significant Ni-rich regions present at the top, centre, and bottom of the weld. This suggests that Ni, having an atomic mass closer to Fe, diffuses more easily into the Fe-rich molten pool regions.

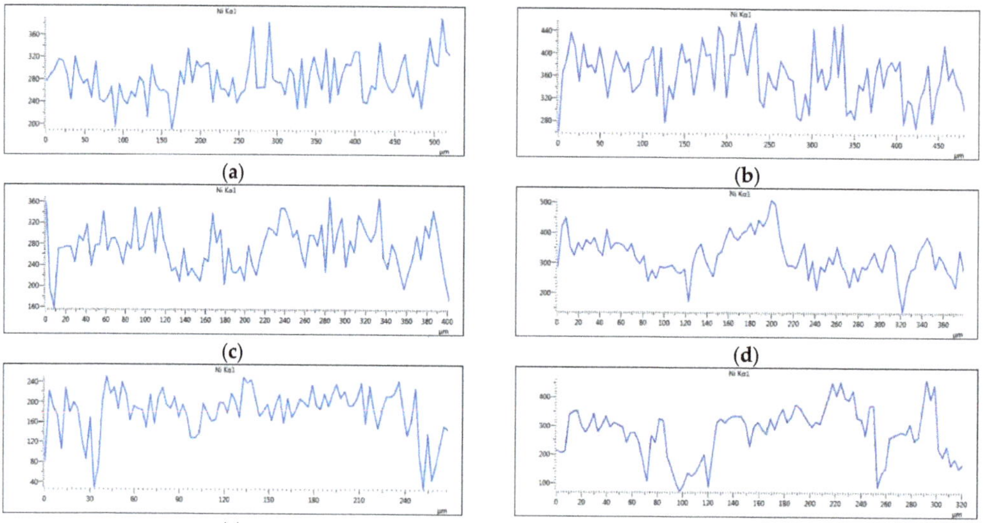

Figure 11. Ni element distribution in different section of welds with different copper–nickel binary: (**a**) top section of 60 μm Cu + 40 μm Ni weld; (**b**) top section of 155 μm Cu + 110 μm Ni weld; (**c**) centre section of 60 μm Cu + 40 μm Ni weld; (**d**) centre section of 155 μm Cu + 110 μm Ni weld; (**e**) bottom section of 60 μm Cu + 40 μm Ni weld; (**f**) bottom section of 155 μm Cu + 110 μm Ni weld.

3.4. Compound and Microstructural Analysis of Welds

Based on the EDS line scan analysis, point scans were conducted on six characteristic regions at the bottom of the weld pool in aluminum–steel laser lap welds with varying copper–nickel coating thicknesses, as shown in Figure 12. Points 1, 2, 3, 5, and 6 were located on the fusion line, while Point 4 was at the bottom of the weld pool's symmetry axis.

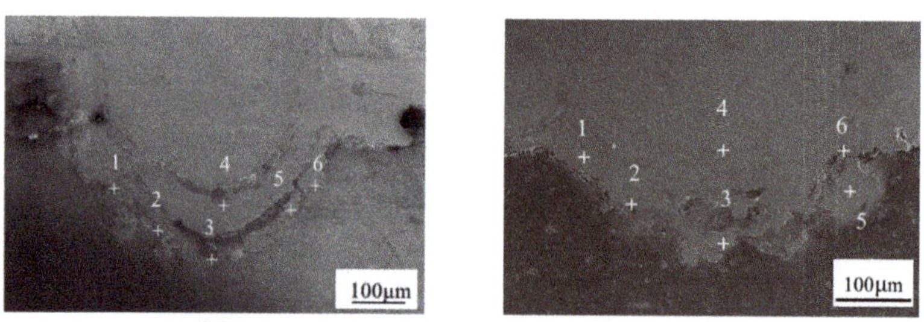

Figure 12. Scanning positions for EDS analysis of welds with different thicknesses of plated layers: (**a**) 60 μm Cu + 40 μm Ni; (**b**) 155 μm Cu + 110 μm Ni.

The atomic percentages of Al, Fe, Cu, and Ni in these regions are shown in Tables 4 and 5. For coatings with 60 μm Cu and 40 μm Ni, the atomic percentages of Fe, Cu, Ni, and Al in regions 1, 2, 4, 5, and 6 were observed to be close to but not exactly 3:1, suggesting that these regions may primarily contain (Fe, Cu, Ni)$_3$Al. However, as noted in Table 4, points 1, 4, and 6 exhibit lower Al content than the theoretical stoichiometric ratio. This discrepancy may be attributed to localized variations in elemental diffusion and the presence of other minor intermetallic phases. Therefore, while these regions likely contain (Fe, Cu, Ni)$_3$Al, the presence of other phases cannot be ruled out. To further clarify the composition, additional point scans and phase identification techniques such as XRD will be considered in future studies. In region 3, the atomic percentage of Fe, Cu, Ni, and Al was approximately 1:3, indicating the presence of (Fe, Cu, Ni)Al$_3$. For coatings with 155 μm Cu and 110 μm Ni, regions 1, 3, 4, 5, and 6 were expected to be dominated by (Fe, Cu, Ni)$_3$Al; however, deviations in Al content were observed, particularly at points 1, 3, and 4 in Table 5. This may be due to the preferential diffusion of Fe and Ni in the molten pool, altering the expected composition. Given these variations, it is possible that some areas contain a mixture of (Fe, Cu, Ni)$_3$ Al and other Fe-rich intermetallic phases rather than a single uniform phase. Future research will focus on a more comprehensive phase analysis to better understand the influence of Cu and Ni additions on phase formation. Meanwhile, region 2 exhibited a Cu:Al ratio of 1:5, suggesting the formation of CuAl$_5$, which further supports the role of Cu in modifying the weld microstructure.

Table 4. EDS compositional analysis of welds at 60 μm Cu + 40 μm Ni (atomic fraction, %).

Placement	Al (at%)	Fe (at%)	Cu (at%)	Ni (at%)	Polymetallic Phase (Chemistry)
1	14.44	68.03	3.42	14.11	(Fe, Cu, Ni)$_3$Al
2	21.05	63.5	3.05	12.4	(Fe, Cu, Ni)$_3$Al
3	76.04	17.31	2.61	4.04	(Fe, Cu, Ni)Al$_3$
4	13.31	70.35	5.95	10.39	(Fe, Cu, Ni)$_3$Al
5	20.11	59.36	8.75	11.78	(Fe, Cu, Ni)$_3$Al
6	12.63	63.58	10.03	13.76	(Fe, Cu, Ni)$_3$Al

Comparisons revealed that at 60 μm Cu + 40 μm Ni, the Cu and Ni layers reacted with Al atoms to form (Fe, Cu, Ni)$_3$Al and similar compounds. This reaction prevented Al atoms from diffusing toward the Fe side, reducing the formation of brittle Fe-Al intermetallic compounds and thus improving weld quality. As the coating thickness increased to 155 μm Cu + 110 μm Ni, the atomic percentages of Cu and Ni in the weld zone further increased. More Cu and Ni atoms reacted with Al atoms, further hindering Al diffusion and enhancing the overall weld quality.

Table 5. EDS compositional analysis of welds at 155 μm Cu + 110 μm Ni (atomic fraction, %).

Placement	Al (at%)	Fe (at%)	Cu (at%)	Ni (at%)	Polymetallic Phase (Chemistry)
1	10.09	61.05	12.12	16.74	(Fe, Cu, Ni)$_3$Al
2	85.81	0	14.05	0.14	CuAl$_5$
3	12.52	45.29	12.95	29.24	(Fe, Cu, Ni)$_3$ Al
4	7.29	57.56	8.13	27.02	(Fe, Cu, Ni)$_3$Al
5	17.25	34.99	26.73	21.03	(Fe, Cu, Ni)$_3$Al
6	20.19	41.70	8.35	29.76	(Fe, Cu, Ni)$_3$Al

4. Conclusions

In this study, on aluminum–steel laser welding, a pre-deposited copper–nickel binary coating was used as an interlayer. Through tensile tests and microscopic analysis, the effects of the copper–nickel binary coating thickness on the mechanical performance, morphology, microstructure, and compound composition of the welds were investigated. The main conclusions are as follows:

1 The pre-deposition of a copper–nickel binary coating between aluminum and steel significantly improved the weld quality in aluminum–steel laser lap welding. It reduced the formation of cracks in the weld zone, resulted in a smoother fusion line interface, and greatly enhanced the mechanical performance of the welds. The weld exhibited optimal shear strength when the Cu and Ni coating thicknesses were 155 μm and 110 μm, respectively, achieving a maximum shear force of 208.09 N, which represents a 56% improvement compared to aluminum–steel lap welding without coatings.
2 As the thickness of the copper–nickel binary coating increased, more Cu and Ni atoms entered the molten pool and reacted metallurgically with Fe and Al atoms, forming multi-element compounds such as $(Fe, Cu, Ni)_3Al$, $(Fe, Cu, Ni)Al_3$, and $CuAl_5$. This effectively suppressed the formation of brittle Fe-Al intermetallic compounds, further enhancing the weld quality.

Author Contributions: Conceptualization, H.Z. and H.G.; methodology, H.Z. and D.M.; investigation, H.Z. and H.G.; resources, D.M.; data curation, H.G.; writing—original draft preparation, H.G.; writing—review and editing, H.Z.; project administration, H.Z.; funding acquisition, H.Z. All authors have read and agreed to the published version of the manuscript.

Funding: This research was funded by "National Natural Science Foundation of China, grant number 51205212" and "Postgraduate Research & Practice Innovation Program of Jiangsu Province grant number KYCX233389" and "Basic Research Program of Nantong grant number JC22022074".

Data Availability Statement: The authors confirm that the data supporting the findings of this study are available within the article.

Conflicts of Interest: The authors declare no conflicts of interest.

References

1. Schimek, M.; Springer, A.; Kaierle, S.; Kracht, D.; Wesling, V. Laser-welded dissimilar steel-aluminum seams for automotive lightweight construction. *Phys. Procedia* **2012**, *39*, 43–50.
2. Wu, X.; Yan, H.; Zhou, Y.; Zhang, P.; Lu, Q.; Shi, H. Review of additive manufactured metallic metamaterials: Design, fabrication, property and application. *Opt. Laser Technol.* **2025**, *182*, 112066.
3. Li, J.; Sun, Q.; Jia, J.; Zhu, W. Research on Pre-Compensation and Shape-Control Optimization of Hemming Structures with Dissimilar Materials Based on Forming Process Chain. *Metals* **2024**, *14*, 1063. [CrossRef]
4. Marumoto, K.; Horai, T.; Morita, D.; Oda, C.; Fujii, T.; Yuzawa, T.; Koba, R.; Yamamoto, M. Development of Hot-Wire Laser Additive Manufacturing for Dissimilar Materials of Stainless Steel/Aluminum Alloys. *J. Manuf. Mater. Process.* **2024**, *8*, 93. [CrossRef]
5. Wallerstein, D.; Salminen, A.; Lusquinos, F.; Comesaa, R.; Pou, J. Recent developments in laser welding of aluminum alloys to steel. *Metals* **2021**, *11*, 622. [CrossRef]
6. Wu, L.; Oliveira, J.; Yang, J.; Xiao, M.; Zheng, M.; Xu, W.; Zhao, Y.; Wang, F.; Zhang, H. Effect of Al-Si Coating on the Interfacial Microstructure and Corrosion Resistance of Dissimilar Laser Al Alloy/22MnB5 Steel Joints. *Metals* **2024**, *14*, 328. [CrossRef]
7. Jiang, F.; Wang, W.; Zhang, X.; Gong, W. Microstructure and Mechanical Properties of Friction Stir Lap Welding Joint of Al/CU Dissimilar Metals. *Metals* **2023**, *13*, 1969. [CrossRef]
8. Shi, X.; Zhang, S.; Li, T.; Meng, X.; Cheng, C.; Pei, J.; Cao, T.; Zhao, J. Corrosion Behavior and Mechanical Property of 5182 Aluminum/DP780 Steel Resistance Spot Welding Joints. *Materials* **2024**, *17*, 2472. [CrossRef]
9. Zhao, Y.; Long, Y.; Li, Z. Research progress of transition layer and filler wire for laser welding of steel and aluminum dissimilar metals. *Int. J. Adv. Manuf. Technol.* **2022**, *119*, 4149.
10. Li, Y.; Liu, Y.; Yang, J. First principle calculations and mechanical properties of the intermetallic compounds in a laser welded steel/aluminum joint. *Opt. Laser Technol.* **2020**, *122*, 105875.

11. Yang, J.; Su, J.; Gao, C.; Zhao, Y.; Liu, H.; Joao, P.; Tan, C.; Yu, Z. Effect of heat input on interfacial microstructure, tensile and bending properties of dissimilar Al/steel lap joints by laser Welding-brazing. *Opt. Laser Technol.* **2021**, *142*, 107218.
12. Yu, J.; Fan, Y.; Zhang, H.; Cao, W.; Zhang, H.; Su, Z.; Gao, J. Interfacial evolution of thermo-compensated resistance diffusion welding of Al/steel joint with Ni interlayer via resistance seam welding. *J. Mater. Eng. Perform.* **2024**, *33*, 6480.
13. Jiang, Q.; Yang, J.; Xiao, R.; Zheng, M.; Jiang, J.; Xue, J.; Xu, W.; Xu, J.; Zhang, H. Outstanding ductility of dissimilar laser Al/steel spot joint using a high entropy alloy interlayer. *Mater. Lett.* **2023**, *349*, 134707.
14. Hu, X.; Xia, C.; Zhang, H.; Wan, G.; Liu, S.; Long, W. The microstructure and mechanical properties of aluminum/steel welded joint with Ni foil interlayers by FSLW. *J. Adhes. Sci. Technol.* **2024**, *38*, 1413.
15. Li, B.; Han, S.; Ao, W.; Yang, Y.; Liu, D. Investigation on microstructure and mechanical properties of the dissimilar lap welding of Al alloy to steel with different filler materials and laser powers. *Int. J. Adv. Manuf. Technol.* **2024**, *134*, 935.
16. Esmaily, H.; Habibolahzadeh, A.; Tajally, M. Improving pulsed laser weldability of duplex stainless steel to 5456 aluminum alloy via friction stir process reinforcing of aluminum by BNi-2 brazing alloy. *Trans. Nonferrous Met. Soc. China* **2019**, *29*, 1401.
17. Yu, G.; Chen, S.; Zhao, Z.; Wen, Z.; Huang, J.; Yang, J.; Chen, S. Comparative study of laser swelding-brazing of aluminum alloy to galvanized steel butted joints using five different filler wires. *Opt. Laser Technol.* **2022**, *147*, 107618.
18. Hu, Y.; Zhang, Y.; Mi, G.; Wang, C.; Zhang, W.; Zhang, X. Effects of Si contents in filling wires on microstructure evolution and properties of Al-steel dissimilar joint by laser welding-brazing. *J. Mater. Res. Technol.* **2021**, *15*, 1896–1904.
19. Shi, Y.; Li, Z.; Yin, C.; Yu, X.; Kou, Y. Effect of alloying elements Cu and Ni on mechanical properties of steel/aluminum laser welded joints. *Optik* **2022**, *255*, 168707.
20. Zhang, Y.; Wang, W.; Li, Z.; Huang, G.; Zhang, H.; Liu, F. Study of the brittleness mechanism of aluminum/steel laser welded joints with copper and vanadium interlayers. *Opt. Laser Technol.* **2023**, *163*, 109319.
21. Li, T.; Zhou, D.; Yan, Y.; Liu, J. Effect of laser-Ti foil coupling on microstructure and mechanical property of steel/aluminum fusion welding joints. *Opt. Laser Technol.* **2021**, *141*, 107114.
22. Guo, W.; Wu, M.; Teng, L.; Liu, L.; Wan, H. Effect of Filler-Wire Composition on Microstructure and Properties of Al/Steel-Welded Joints by Laser Welding–Brazing. *Crystals* **2025**, *15*, 266. [CrossRef]

crystals

MDPI

Article

Dry Sliding Wear Behavior and Mild–Severe Wear Transition of the AA2195-T6 Alloy under Different Loads

Qingqiang Chen [1,2,*], Yalei Yu [2], Guanjie Ma [2], Xingzi Sun [2] and Laixiao Lu [2]

1 Key Laboratory of Electromagnetic Processing of Materials, Ministry of Education, Northeastern University, Shenyang 110819, China
2 School of Mechanical and Electronic Engineering, Shandong Jianzhu University, Jinan 250101, China
* Correspondence: 13770@sdjzu.edu.cn

Abstract: The mild–severe wear transition of aluminum alloys is considered evidence that the wear changes from a stable state to an unstable state, which is of great importance in engineering applications. The purpose of this study is to evaluate the mild–severe wear transition of the 2195 Al–Li alloy for different loads and to elucidate the causes behind it. To this end, dry sliding tribometric tests were carried out by varying the normal load from 2 to 40 N at room temperature. The results show that the change in wear rate can be divided into three distinct stages, including weak growth at low load (2–4 N), rapidly increased growth at medium load (8–16 N), and gradually increased growth at high load (32–40 N). The transition from mild to severe wear is observed at loads ranging from 4 to 8 N. Characterization of the worn surface of the Al–Li alloy via scanning electron microscopy shows that abrasion and oxidation are the dominant wear phenomena in the mild wear regime. On the other hand, delamination, adhesion, and severe plastic deformation become dominant in the severe wear regime. The reason for the occurrence of the transition is the tribo-induced plastic deformation of the substrate.

Keywords: Al–Li; tribological behavior; spray forming; wear phenomenon

1. Introduction

Due to their remarkable properties, aluminum–lithium (Al–Li) alloys have become the preferred material for aerospace, automotive, and commercial applications. Their high elastic modulus stiffness, low density, high specific strength, and excellent cryogenic properties make them superior to other materials. Lithium is the lightest metal (0.534 g/cm^3) and dissolves readily in aluminum. It remarkably reduces the alloy density by 3% while raising its elastic modulus by 6% for every 1 wt% addition [1]. However, two of the main limitations in the application of Al–Li alloys are friction and wear, which affect their performance and service life. To improve the wear resistance of components, it is essential to clarify the tribological behavior of the metal under different conditions [2–4]. In this context, material scientists have reported several studies on improving the tribological behavior of Al-based alloys. For example, Schuette et al. [5] prepared thick Al–Si alloy coatings with different Si contents via surfacing. Furthermore, they studied the mechanisms that improved the wear resistance of Si-rich Al alloy coatings. Yang et al. [6] investigated the wear behavior of Al–Zn–Mg–Cu alloys under various loads and temperatures. Tarasov et al. [7] focused on the direct and reverse adhesion transfer between Al–Mg alloys and bearing steel specimens during sliding contact at high temperatures. In addition, many studies have focused on studying the friction and wear behavior of aluminum matrix composites [8].

First- and second-generation Al–Li alloys are lightweight alloys extensively used in several applications due to their high specific strength, high specific stiffness, and low density. Similarly, the new third-generation Al–Li alloys exhibit improved toughness and ductility due to the increased Cu/Li content ratio. Therefore, these enhanced Al–Li alloys have been widely used in manufacturing aerospace components as substitutes for 2xxx and 7xxx series aluminum alloys. Among third-generation Al–Li alloys, the 2195 Al–Li alloy has gained increased attraction. For 2195 Al–Li alloys, most of the published research has focused on how manufacturing technology affects the material microstructure and performance. For example, Hatamleh et al. [9] thoroughly studied the fatigue crack growth performance of a friction stir welded 2195 Al–Li alloy, which was subjected to different surface treatment technologies. Nayan et al. [10] studied the alloy microstructure evolution by conducting a hot isothermal plane-strain compression test. Wang et al. [11] investigated the impact of aging on the intergranular corrosion, exfoliation corrosion, and stress corrosion cracking behavior of a 2195 Al–Li alloy.

Although wear is not the primary factor affecting the service life of aerospace materials, structural components, such as seat rails and landing gear axles, are still affected by friction and wear problems. Therefore, the friction and wear behavior of Al–Li alloys have attracted the growing attention of researchers in recent years. For instance, using plasma electrolytic oxidation, Cheng et al. [12] prepared wear-resistant coatings containing relatively large amounts of α-Al_2O_3 on a 2A97 Al–Cu–Li alloy. They studied the effect of the $NaAlO_2$ electrolyte concentration on wear resistance. Li et al. [13] investigated the tribological behavior of a 2297 Al–Cu–Li alloy under various loads and sliding speeds using a ball-on-disk configuration. They analyzed the wear characteristics, including the wear coefficient, wear rate, and morphology of the worn surface and debris. In addition, friction and wear must be considered in the plastic forming of metal materials, surface treatment, and other production processes [4,14].

The concept of mild and severe wear was first proposed by Archard and Hirst for aluminum alloys [15]. This concept has now been extended to magnesium, steel, and other alloys [16]. In the mild wear regime, the wear is stable and the wear rate is low. By contrast, in the severe wear regime, the wear is unstable and the wear rate is high. Therefore, it is of great significance to study the mild–severe wear transition of aluminum alloys for promoting their tribological application. Therefore, the mild–severe wear transition behavior of various alloys, such as Al–Si alloys [17], Al–Zn alloys [6], Al–Sn alloys [18], and aluminum matrix composites [19], has been extensively studied in previous works.

Spray forming is an advanced solidification technology for developing large-scale ingots with a uniform microstructure and without the macro-segregation of chemical components [20]. As a result, numerous popular alloys, especially Al–Li alloys, have been produced via spray forming. In the presented study, spray forming produced a 2195 Al–Li alloy ingot. The ingot was extruded to eliminate microstructural defects, such as micropores. The effect of the load on the alloy friction coefficient and wear rate was studied in detail. The mild–severe wear transition was evaluated. The wear morphologies and friction-induced plastic deformation structures were characterized.

2. Experimental Procedures

2.1. Materials

The test samples were extracted from the spray-formed 2195 Al–Li alloy billet. The chemical composition of the billet was Al–3.78 Cu–0.90 Li–0.56 Mg–0.34 Ag–0.11 Zr. Figure 1a shows the schematics of the spray forming technique used in this study. The raw materials were aluminum (99.996% purity), copper (99.99% purity), silver (99.99% purity), Al–10 Li, Al–10 Zr, and Al–50 Mg master alloys. An Osprey spray casting system with a 2000 kg aluminum melt capacity was used. The raw materials were melted in an inductively heated crucible in an Ar inert atmosphere. At 750 °C, the melt was fed into a two-stage atomizer and atomized by an argon jet with a high flow rate and pressure. The spray was collected on a rotating steel collecting plate and gradually solidified into a preform. The rotation rate of the collecting plate was 40 rpm, and the deposition rate was 15 kg/min. The dimensions of the preformed billet were 650 mm in diameter and 1800 mm in length. No macroscopic defects were found in the plate, such as thermal tearing or rough shrinkage.

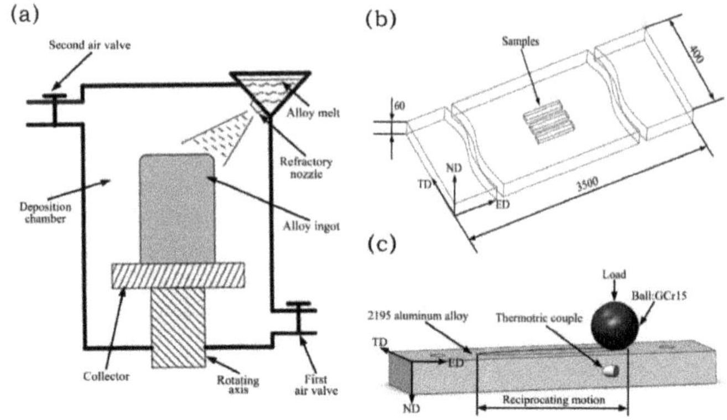

Figure 1. Schematics of (**a**) The spray forming process, (**b**) Extruded plate and sample position, and (**c**) Friction and wear experiment.

2.2. Processing

After forming, the billet was homogenized at 450 °C for 10 h. The billet was machined to a diameter of 490 mm and a length of 1000 mm. Machining removed surface defects and transformed the billet according to the capacity of the extrusion cylinder. Hot extrusion was conducted using a horizontal extruder (80 MN (meganewton), Taiyuan Heavy Industry CO., LTD., Taiyuan, China) to obtain the plate with a cross-section size of 400 mm × 60 mm. The diameter of the extrusion cylinder was 500 mm. The extrusion ratio was 8.17, and the dummy block speed was 0.1 mm/s. The extrusion cylinder and mold temperatures were 460 °C and 420 °C, respectively. After extrusion, the samples were cut by wire-electrode cutting. To avoid the influence of microstructure heterogeneity, all samples were extracted from the center of the plate. The sampling position and its relationship with the extrusion direction are shown in Figure 1b. The samples were solution treated before testing at a heating rate of 4.25 °C/min. The temperature was kept at 510 °C for 2 h. At the end of the solution treatment, water quenching was performed at 60 °C. Then, the samples were aged at 170 °C for 28 h (T6) and cooled in air.

2.3. Analysis and Characterization

Optical microscopy (OM, DMI5000M, Leica Microsystems, Weitzlar, Germany) was used to analyze the morphology of the extruded plate. The OM samples were mechanically polished with different sandpapers and flannels. A cotton ball dipped in Kellar's reagent (1 mL HF + 1.5 mL HCl + 2.5 mL HNO_3 + 95 mL H_2O) was used to wipe the polished surface for 35 s. The morphology and type of the second phase were identified via transmission electron microscopy (TEM, HT-7700, Hitachi High-tech Company, Tokyo, Japan). The acceleration voltage was 200 kV. For the TEM analysis, a plasma ion polisher (Model 691, Gatan, CA, USA) was used to prepare thin foils with a thickness of 80 nm. The sliding dry friction and wear properties of the alloy were tested using a reciprocating friction and wear testing machine (MWF-05, Shandong Baohang Machine Equipment Manufacturing Co., Ltd., Jinan, China).

The ball-on-disk wear tests were conducted under dry conditions at a standard temperature (~25 °C). A high carbon chromium-bearing steel ball (GCr15, φ6 mm) was used because of its high hardness (~63 HRC) and the fact that it can concentrate a significant amount of wear on the alloy surface. A wide range of loads was applied during the pretest to thoroughly evaluate the Al–Li alloy wear phenomena. However, it is difficult to accurately predict the wear rate under small loads, whereas excessively high loads cause the instability of the friction process. Based on the tests, the optimal experimental load range was 2–40 N. The stroke and frequency were kept constant at 15 mm and 2.5 Hz, respectively. The sliding distance was 55 m, and the sliding direction was parallel to the extrusion direction. The positional relationship between the friction contact surface and the extruded plate is shown in Figure 1c. The samples were cleaned with alcohol and distilled water and dried in chilled air.

After the wear test, the three-dimensional (3D) topographic images of the worn surface were acquired via laser scanning confocal microscopy (LSM900, ZEISS, Oberkochen, Germany). The volume V (mm^3) of the worn-out material was calculated using the software provided by the confocal measurement system. The specific wear rate w ($mm^3 \cdot m^{-1}$) was calculated using the following equation:

$$w = V/s \tag{1}$$

where s (m) is the total sliding distance. The friction tests were repeated six times under each loading condition and then averaged to obtain the final result. The standard deviation was calculated to determine the dispersion of the data. To analyze the wear phenomenon, the morphology of the sliding track was observed via scanning electron microscopy (SEM, Quant FEG 250FEI, FEI, Hillsboro, OR, USA) combined with energy-dispersive spectroscopy (EDS). SEM and EDS were also utilized to observe the debris morphology of the samples. Finally, the worn sample was cut parallel to the sliding direction. After mechanical polishing and chemical corrosion, OM was used to observe the microstructure of the friction-induced deformed layer.

3. Results and Discussion

3.1. Microstructure Analysis before Friction

The microstructure behavior of the 2195 Al–Li plate after the heat treatment is shown in Figure 2. It can be observed that the aluminum plate was fully recrystallized with marginally elongated grains. The size of the recrystallized grains ranges from 5 to 12 μm. TEM was used to analyze the morphology and the precipitated phase. The results are shown in Figure 3. Figure 3a shows the brightfield (BF) images of the alloy microstructure. Figure 3b shows the selected area electron diffraction (SAED) pattern. It can be observed that there are three types of fine second phases in the alloy, namely the needle-like phases T1(Al_2CuLi) and θ′(Al_2Cu), and the spherical phase δ′(Al_3Li). Similar findings have been reported by Yang et al. [21,22].

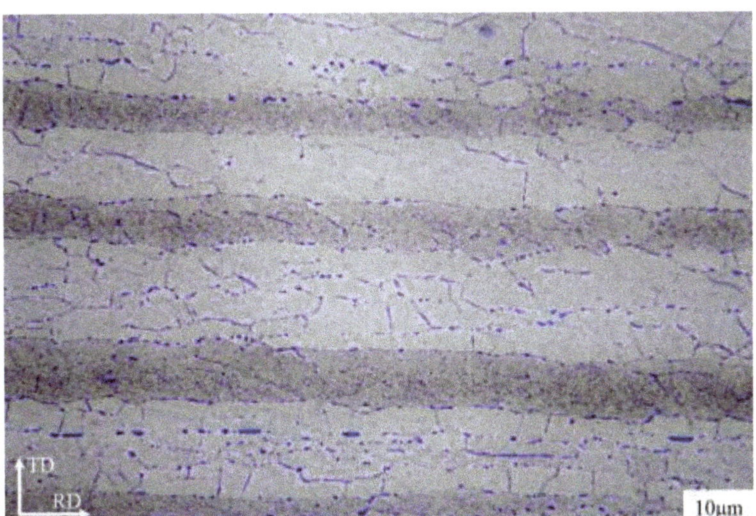

Figure 2. OM image of the 2195 Al–Li alloy.

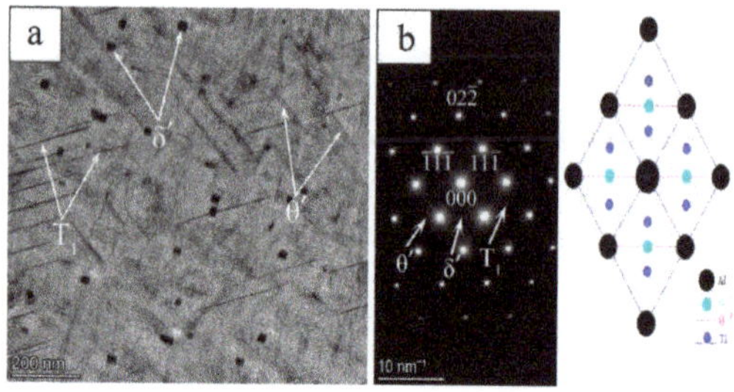

Figure 3. TEM microstructure of the 2195 Al–Li extruded plate: (**a**) BF image and (**b**) Corresponding SAED pattern.

3.2. Results of the Friction and Wear Experiment

3.2.1. Coefficient of Friction and Wear Rate

The friction coefficient and wear rate of the 2195 Al–Li alloy as a function of the applied load are shown in Figure 4. In Figure 4a, the wear rate of the alloy shows three different stages: the stage in which it increases slowly at a low load (Stage I, 2−4 N), the stage in which it increases rapidly at a medium load (Stage II, 8–16 N), and the final stage in which it increases gradually to the highest level (Stage III, 32−40 N). At 2−4 N, the wear rate is about 0.1 $mm^3 \cdot m^{-1}$. When the load increases from 4 to 16 N, although the load rises by a factor of 3, the wear rate increases by a factor of 10 (0.96 $mm^3 \cdot m^{-1}$). With increasing load, the increase in the wear rate is reduced. At 40 N, the wear rate reaches its maximum value of 1.34. Figure 4b,c show that the friction coefficient of the alloy is minimal and stable in Stage 1 (2−4 N), about 0.1. However, during the second and third stages (8−40 N), the friction coefficient rises rapidly to about 0.6. At the same time, the friction coefficient fluctuates dramatically, which means the wear process becomes unstable.

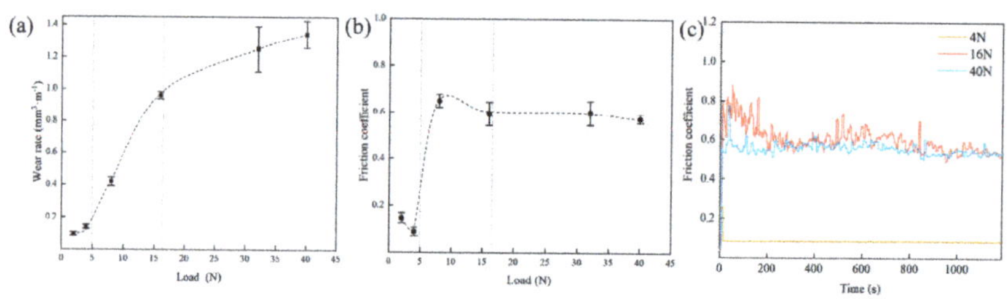

Figure 4. (**a**) Wear rate and (**b**) Friction coefficient of the 2195 Al–Li alloy under different loads; (**c**) Friction coefficient as a function of time.

The variation in the friction coefficient and wear rate as a function of the applied load indicates that the alloy is in the mild wear regime in Stage 1 and enters the severe wear regime in Stages 2 and 3. The mild–severe wear transition has also been observed in other alloys [23,24].

3.2.2. Wear Phenomenon

The 3D profilometry analysis results of the standard samples are shown in Figure 5. For simplicity, only the samples tested at 4 N (Stage I), 16 N (Stage II), and 40 N (Stage III) were selected as the representatives. The color change shown in the figure corresponds to the variation in the wear track depth. The line profile of the surfaces shows that the width and depth of the abrasion mark increase exponentially with the applied load. At 4 N, the width of the abrasion mark is 238 μm, and the depth is 0.5 μm. However, when the load increases to 16 N, the width and depth rapidly increase to 991 and 58 μm, respectively. By increasing the load to 40 N, the width and depth of the abrasion mark reach 2967 and

357 μm, respectively. This change indicates that the increased load causes the sliding ball to push away more metal, which causes an increase in the sliding resistance and friction coefficient. The significant difference in wear volume is related to the wear behavior of metals under different loads. The wear surface morphology was observed using SEM under loads of 2–40 N for a comprehensive understanding, as shown in Figure 6a–f.

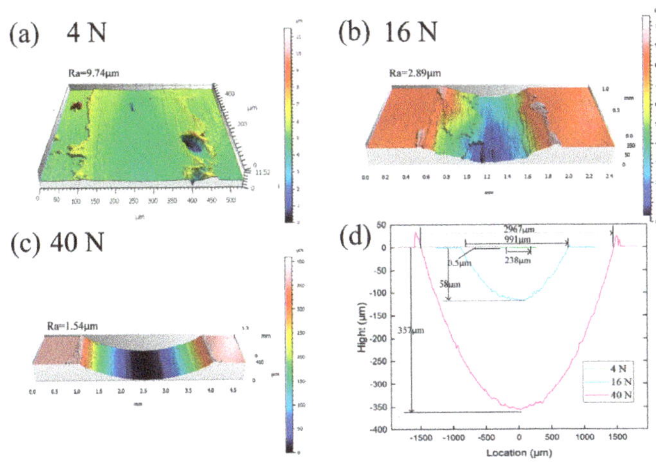

Figure 5. (**a–c**) 3D wear track surface profiles of the 2195 Al–Li alloy and (**d**) Typical cross-sectional profiles of the wear tracks. The sliding direction is perpendicular to the screen.

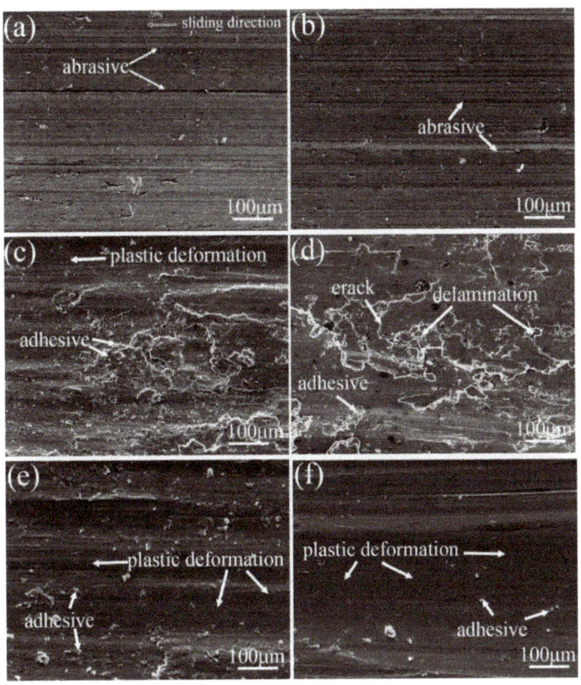

Figure 6. Morphology of the worn surface: (**a**) 2 N, (**b**) 4 N, (**c**) 8 N, (**d**) 16 N, (**e**) 32 N, and (**f**) 40 N.

Stage 1: Due to the abrasive wear, grooves and scratches can be observed on the matrix for an experimental load of 2−4 N (Figure 6a,b). The figure shows that the morphology is affected by wear debris particles formed during friction and hard microscopic bumps on the paired ball surface [25]. Suresha et al. [26] explained that the size of the micro-bumps on the surface of the tribo-pair material influences the wear rate more than the applied load. Considering this effect, it can be assumed that in the present study the wear rate of the 2195 Al−Li alloy increases slowly in Stage 1.

The oxygen composition of the worn surface was analyzed via EDS. This is shown in Figure 7. It can be observed that at 2 and 4 N, an excessive amount of elemental oxygen appears on the surface of the alloy after being worn (Table 1). This phenomenon indicates that an oxide film is formed on the metal surface during the friction of the alloy. Then, the oxide film instantly breaks and generates debris under the action of cyclic contact stress. The continuous formation and detachment of the oxide film are regarded as oxidation wear [27]. Many models have been proposed to explain the oxidative wear of metals [28,29]. These models use different parameters and conditions; however, these are based on the same assumption that the behavior of the oxide film affects the oxidative wear process. Quinn [30] proposed the oxidation wear model of steel considering the light wear stage:

$$\omega_L = WAdexp(-Q/RT)/\rho^2 V p_m \xi^2 \tag{2}$$

where ω_L represents the wear rate, ξ is the maximum thickness of the accumulating oxide film, d is the relative sliding distance between the two metals, W is the applied load, A is the rate constant of Arrhenius's empirical formula, Q is the activation energy of the metal oxidation reaction, R is the ideal gas constant, T is the temperature of the contact surfaces, ρ is the oxide density, V is the sliding velocity, and p_m is the yield pressure of the softer metal. From Equation (2), it can be inferred that when the 2195 Al–Li alloy is in Stage 1, the material loss caused by oxidative wear is positively correlated with the load W and the temperature T. An increase in the applied load results not only in an increase in W but also in an increase in T because of the accumulation of friction heat. Therefore, in Stage 1, an increased loading can increase the contribution of oxidative wear.

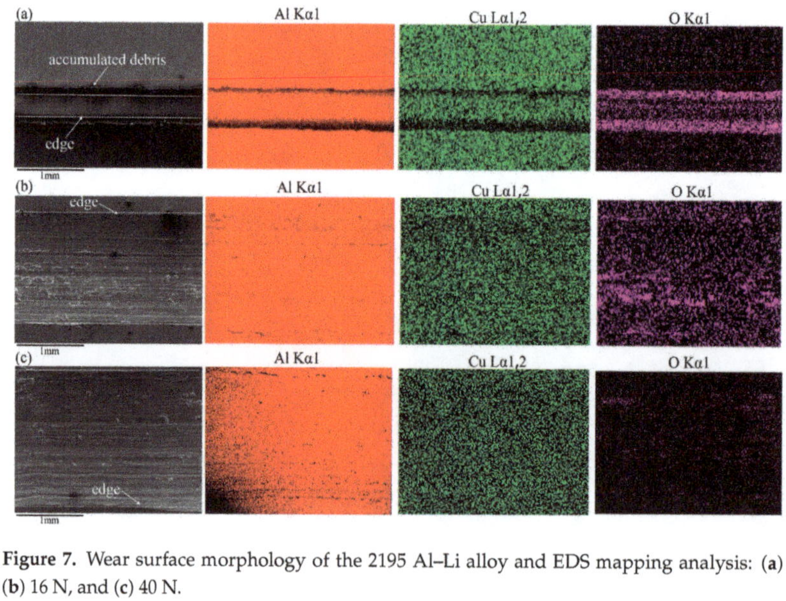

Figure 7. Wear surface morphology of the 2195 Al–Li alloy and EDS mapping analysis: (**a**) 4 N, (**b**) 16 N, and (**c**) 40 N.

Table 1. EDS surface scanning analysis results (wt%).

Element Content (wt%)	Al	Cu	O
Tested at 4 N	91.18	4.55	4.27
Tested at 16 N	93.29	4.71	2.00
Tested at 40 N	94.02	4.68	1.30

Stage 2: Under the normal loads of 8 and 16 N, the wear phenomenon of the alloy was the same. The SEM images shown in Figure 6c,d indicate the presence of patches of re-deposited material and adhesive wear. In the adhesion and sliding processes, strong ionic and covalent bonds form between the micro-convex bodies of the friction pairs, increasing the friction coefficients. The wear volume can be calculated according to the following expression [31]:

$$V = \frac{1}{3} \times \left(\frac{n}{\lambda}\right) \times \left(\frac{b}{a}\right)^3 \times \left(\frac{P \times \tau}{P_m}\right) \quad (3)$$

where V is the wear volume, n is the number of wear elements generated in the microscopic contact area where adhesion occurs, and a and b are the mean radius of the microscopic bumps of the wear materials and tribo-paired materials, respectively. P is the applied load, and P_m is the yield stress of the wear material (the softer material). τ is the sliding distance, and λ is a constant determined by the chemisorption capacity of the surrounding molecules. From the wear volume expression, it is clear that the volume of material loss caused by adhesive wear is proportional to the applied load and sliding distance. Therefore, when the wear of the 2195 Al–Li alloy enters Stage 2, the contribution of adhesion to the wear may increase with increasing load.

The area outside the adhesive wear is smooth, indicating that the metal undergoes plastic deformation under the combined effect of a high load and frictional heat. The continuously accumulating heat at the contact surface causes thermal softening, increasing the alloy plasticity and generating a smooth surface. This is referred to as severe plastic deformation [32]. The wear phenomenon makes the alloy prone to plastic deformation and failure when subjected to micro-plowing. As a result, the surface alloy is pulled out of the friction contact surface and is constantly detached, forming large lamellar debris. Zheng et al. [33] studied the surface microstructure of an Al–Zn–Mg–Cu alloy subjected to friction and wear using TEM. The results show that severe plastic deformation causes dense dislocations in the cell walls. Later, micro-lamellae with a high dislocation density are formed, and the equiaxed dislocation cells develop inside the micro-lamellae. Finally, the lamellae are gradually separated into randomly oriented ultrafine grains. Published studies have explained that similar phenomena occur in Inconel and magnesium alloys [34,35].

In addition, spalling and cracking perpendicular to the sliding direction occur under the applied load, which means that the surface layer experiences delamination wear. According to the Hertz contact theory, the maximum shear stress occurs at a certain depth below the surface during the sliding process of the ball [36]. The stress causing plastic deformation in the substrate induces the formation of many dislocations and defects. Greiner et al. elucidated this process via systematic model experiments and discrete dislocation dynamics simulations [37]. This phenomenon is repeated as the sliding proceeds, gradually forming fatigue microcracks and holes in the alloy surface [38].

The EDS analysis reveals that the O element content decreased significantly (Table 1). This behavior can be attributed to the appearance of adhesive and delamination wear, which increase the rate of surface wear failure. As a result, the oxide layer is removed faster than it can form. Different from the low-load conditions (2–4 N), the transformation of the wear phenomenon is a fundamental reason for the rapid increase in the wear rate and the mild–severe wear transition.

Stage 3: A few areas demonstrate adhesive wear characteristics when the load exceeds 30 N (Figure 6e,f). However, most areas of the friction trace become smooth. The results of the surface roughness analysis also confirm this phenomenon (Figure 5). In this stage, the material loss rate is much higher than the oxide film growth rate. The oxygen element content on the alloy-worn surface is reduced to the minimum, as shown in Table 1 and Figure 7. Severe plastic deformation mechanisms dominate the wear of the alloy.

Figure 8 shows the morphology of the wear debris of the 2195 Al–Li alloy under different loads. The samples tested at 4 N (Stage I), 16 N (Stage II), and 40 N (Stage III) were chosen as the representatives. When the experimental load was 4 N, it was observed that the average size of the debris was small. Most of the debris was in the form of fine granular and irregular lumps (Figure 8a). The debris diameter was between 10 and 100 μm. At 16 N, the wear entered the second and third stages. At the same time, the shape of the debris also changed to large lamellae and bands (Figure 9b). It is commonly assumed that fine granular and irregular lumpy debris is produced from abrasive wear. The primary sources of the band-like and lamellar debris are the delamination wear, adhesive wear, and severe plastic deformation mechanisms. Most debris is more than 200 μm in diameter. At 40 N, the shape of the chip does not change appreciably, but the size of the debris increases marginally.

Figure 8. Wear debris morphology of the 2195 Al–Li alloy: (**a**) 4 N, (**b**) 16 N, and (**c**) 40 N.

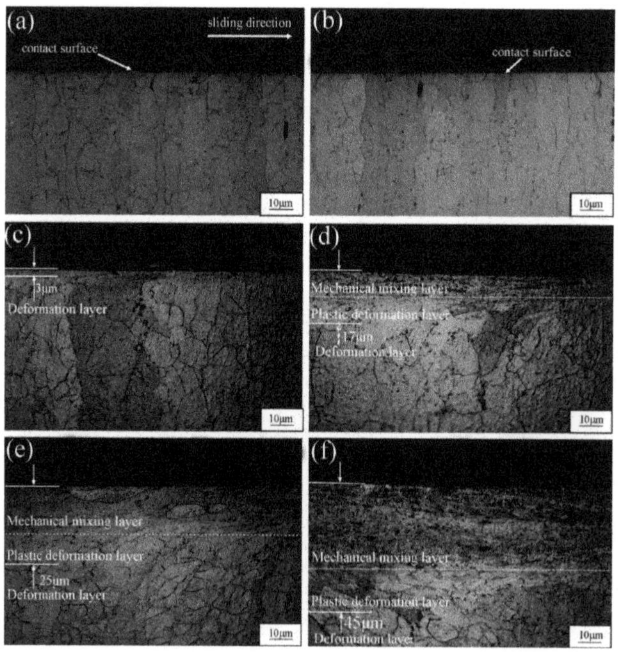

Figure 9. Microstructure of the region near the wear mark: (**a**) 2 N, (**b**) 4 N, (**c**) 8 N, (**d**) 16 N, (**e**) 32 N, and (**f**) 40 N.

3.2.3. Plastic Deformation Behavior Induced by Friction

The microstructural evolution of the aluminum alloy during sliding is considered the main factor that changes during the wear phenomenon, determining the wear properties and mild–severe wear transition. Figure 9 shows the microstructural changes of the substrate as a function of the load. Almost no trace of plastic deformation is observed in Stage 1 (Figure 9a,b). After raising the load (Stages 2 and 3), evident plastic deformation occurred in the substrate. The load increases from 8 to 16 N, increasing the plastic deformation zone depth from 3 to 17 μm, as shown in Figure 9a,b. At the same time, layered structures appear in the deformation zone. The plastic deformation of the metal immediately adjacent to the friction contact surface is exceptionally high, resulting in unrecognizable grain boundaries. The zone immediately adjacent to the friction contact surface is usually referred to as the mechanical mixing layer (MML) and is about 10 μm thick. At 32 N, the depths of the deformation zone and the MML increase simultaneously. The thicknesses of the deformation layer and the MML are 25 and 15 μm, respectively. At the same time, cracks extending from the inner metal to the surface appear in the MML. As discussed in Section 3.2.2, crack propagation causes the spalling of the metal, resulting in delamination wear and spalling pits. This behavior is shown in Figure 6. At 40 N, the plastic deformation of the metal increases, and the thickness of the MML region is nearly doubled. The thickness of the deformation layer increases to 45 μm, and the thickness of the MML layer becomes 30 μm.

During sliding, noticeable plastic deformation occurs in the friction contact area of the metal, which changes the microstructure of the surrounding wear marks. This phenomenon has attracted increasing attention from the research community. Studies on Al–Cu and Al–Mg alloys have shown that the resistance of the metal to the formation and shedding of the MML layer during friction determines the wear behavior [39,40]. Chen et al. investigated the friction-induced deformation behavior of Cu–Al alloys. They found that the transition of the dynamic recrystallization structure to the outermost MML plays a vital role in the tribological and wear properties [41]. For Al–Li alloys, the change in microstructure leads to a change in the wear phenomenon from abrasive to layered wear and finally leads to the transition from mild to severe wear. Therefore, if the degree of plastic deformation caused by friction can be reduced via aging strengthening or other strategies, severe wear may be avoided. This is conducive to the improvement in the wear resistance of the alloy.

4. Conclusions

Evaluating the reasons for the mild–severe wear transition is a prerequisite for delaying the occurrence of such a transition. In this work, the mild–severe wear transition of the 2195 Al–Li alloy for different loads was investigated. The analysis in this study shows that controlling the tribo-induced plastic deformation of the 2195 Al–Li alloy should be an important way to delay the occurrence of the mild–severe wear transition. This provides a breakthrough for future research on improving the wear resistance of Al–Li alloys.

The main conclusions of this work are as follows:

1. When the load increases from 2 to 40 N, the change in the wear rate can be divided into three stages. In Stage 1 (2–4 N), the wear rate does not change significantly with increasing load. The alloy is in the mild wear regime. In Stages 2 (8–16 N) and 3 (32–40 N), the rate at which the wear rate increases firstly rises rapidly and then is gradually reduced. The alloy is in the severe wear regime;
2. The main wear phenomena are abrasion, adhesion, oxidation, delamination, and severe plastic deformation. When the load ranges from 2 to 4 N, the wear is dominated by abrasion and oxidation. At 8 and 16 N, the wear phenomenon changes to adhesion, delamination, and severe plastic deformation. When the load exceeds 32 N, the dominant wear phenomenon is severe plastic deformation;
3. The change in the wear phenomenon is affected by the microstructure of the substrate. When the load exceeds 8 N, friction and wear cause plastic deformation of the 2195 Al–Li alloy. The deformation results in the mechanical mixing and the formation of

plastic deformation zones near the wear contact surface. By increasing the load, the thickness of the plastic deformation layer rises rapidly.

Author Contributions: Conceptualization, Q.C.; Methodology, Q.C.; Investigation, Q.C., G.M. and X.S.; Resources, Y.Y.; Data curation, Y.Y. and L.L.; Writing–original draft, Y.Y.; Writing–review & editing, Q.C. All authors have read and agreed to the published version of the manuscript.

Funding: This work was supported by Project supported by the Natural Science Foundation of Shandong Provincial, China (ZR2020QE161), the Key Laboratory of Electromagnetic Processing of Materials, Ministry of Education, Northeastern University (NEU-EPM-016).

Data Availability Statement: Not applicable.

Conflicts of Interest: We declare we have no commercial or associative interest representing a conflict of interest in connection with the work submitted.

References

1. Zhang, P.; Chen, M. Progress in characterization methods for thermoplastic deforming constitutive models of Al–Li alloys: A review. *J. Mater. Sci.* **2020**, *55*, 9828–9847. [CrossRef]
2. Samal, P.; Vundavilli, P.R.; Meher, A.; Mahapatra, M.M. Recent progress in aluminum metal matrix composites: A review on processing, mechanical and wear properties. *J. Manuf. Process.* **2020**, *59*, 131–152. [CrossRef]
3. Ramezani, M.; Neitzert, T.; Pasang, T.; Sellès, M.A. Frictional Properties of AZ80 and ZE10 Magnesium Alloys under Dry and Lubricated Contact Conditions. *Procedia Eng.* **2014**, *81*, 1836–1841. [CrossRef]
4. Ilie, F. Effect of the Etching on Chemical Mechanical Planarization of the Selective Layer Surface. *Int. J. Mater. Sci. Appl.* **2017**, *6*, 193. [CrossRef]
5. Schütte, M.R.; Ehrich, J.; Linsler, D.; Hanke, S. Effects of Microstructure Modification by Friction Surfacing on Wear Behavior of Al Alloys with Different Si Contents. *Materials* **2022**, *15*, 1641. [CrossRef] [PubMed]
6. Yang, Z.-R.; Sun, Y.; Li, X.-X.; Wang, S.-Q.; Mao, T.-J. Dry sliding wear performance of 7075 Al alloy under different temperatures and load conditions. *Rare Met.* **2015**, *41*, 1057–1062. [CrossRef]
7. Tarasov, S.Y.; Filippov, A.; Kolubaev, E.; Kalashnikova, T. Adhesion transfer in sliding a steel ball against an aluminum alloy. *Tribol. Int.* **2017**, *115*, 191–198. [CrossRef]
8. Singh, J. Fabrication characteristics and tribological behavior of Al/SiC/Gr hybrid aluminum matrix composites: A review. *Friction* **2016**, *4*, 191–207. [CrossRef]
9. Hatamleh, O. A comprehensive investigation on the effects of laser and shot peening on fatigue crack growth in friction stir welded AA 2195 joints. *Int. J. Fatigue* **2009**, *31*, 974–988. [CrossRef]
10. Nayan, N.; Gurao, N.P.; Murty, S.N.; Jha, A.K.; Pant, B.; Sharma, S.; George, K.M. Microstructure and micro-texture evolution during large strain deformation of an aluminium–copper–lithium alloy AA 2195. *Mater. Des.* **2015**, *65*, 862–868. [CrossRef]
11. Wang, X.-H.; Wang, J.-H.; Yue, X.; Gao, Y. Effect of aging treatment on the exfoliation corrosion and stress corrosion cracking behaviors of 2195 Al–Li alloy. *Mater. Des.* **2015**, *67*, 596–605. [CrossRef]
12. Cheng, Y.-L.; Cao, J.-H.; Mao, M.-K.; Peng, Z.-M.; Skeldon, P.; Thompson, G. High growth rate, wear resistant coatings on an Al–Cu–Li alloy by plasma electrolytic oxidation in concentrated aluminate electrolytes. *Surf. Coat. Technol.* **2015**, *269*, 74–82. [CrossRef]
13. Li, G.; Hao, S.; Gao, W.; Lu, Z. The Effect of Applied Load and Rotation Speed on Wear Characteristics of Al-Cu-Li Alloy. *J. Mater. Eng. Perform.* **2022**, *31*, 5875–5885. [CrossRef]
14. Davoudi, M.; Nejad, A.F.; Koloor, S.S.R.; Petrů, M. Investigation of effective geometrical parameters on wear of hot forging die. *J. Mater. Res. Technol.* **2021**, *15*, 5221–5231. [CrossRef]
15. Archard, J.F.; Hirst, W. The wear of metals under unlubricated conditions. *Proc. R. Soc. London. Ser. A Math. Phys. Sci.* **1956**, *236*, 397–410. [CrossRef]
16. Yin, C.-H.; Liang, Y.-L.; Liang, Y.; Li, W.; Yang, M. Formation of a self-lubricating layer by oxidation and solid-state amorphization of nano-lamellar microstructures during dry sliding wear tests. *Acta Mater.* **2018**, *166*, 208–220. [CrossRef]
17. Banerji, A.; Edrisy, A.; Francis, V.; Alpas, A. Effect of bio-fuel (E85) addition on lubricated sliding wear mechanisms of a eutectic Al–Si alloy. *Wear* **2014**, *311*, 1–13. [CrossRef]
18. Wu, X.; Wang, D.; De Andrade, V.; Jiang, Y.; Wang, W.; Wen, S.; Gao, K.; Huang, H.; Chen, S.; Nie, Z. Dry sliding wear of microalloyed Er-containing Al–10Sn–4Si–1Cu alloy. *J. Mater. Res. Technol.* **2020**, *9*, 14828–14840. [CrossRef]
19. Al-Qutub, A.; Khalil, A.; Saheb, N.; Hakeem, A. Wear and friction behavior of Al6061 alloy reinforced with carbon nanotubes. *Wear* **2012**, *297*, 752–761. [CrossRef]
20. Zhang, Q.; Zhang, C.; Lin, J.; Zhao, G.; Chen, L.; Zhang, H. Microstructure analysis and low-cycle fatigue behavior of spray-formed Al–Li alloy 2195 extruded plate. *Mater. Sci. Eng. A* **2018**, *742*, 773–787. [CrossRef]
21. Yang, Y.; Lian, X.; Zhou, K.; Li, G. Effects of laser shock peening on microstructures and properties of 2195 Al-Li alloy. *J. Alloys Compd.* **2018**, *781*, 330–336. [CrossRef]

22. Yang, Y.; Zhou, K.; Li, G. Surface gradient microstructural characteristics and evolution mechanism of 2195 aluminum lithium alloy induced by laser shock peening. *Opt. Laser Technol.* **2018**, *109*, 1–7. [CrossRef]
23. Prabhu, M.S.; Perumal, A.E.; Arulvel, S. Development of multi-pass processed AA6082/SiCp surface composite using friction stir processing and its mechanical and tribology characterization. *Surf. Coat. Technol.* **2020**, *394*, 125900. [CrossRef]
24. Chen, Q.; Yu, Y.; Sun, J.; Jing, C.; Zhao, Y.; Wang, J. Investigation of the Wear Behavior of Surface Welding AZ91 and AZ91+Gd Alloys under Variable Loading Conditions. *Crystals* **2021**, *11*, 554. [CrossRef]
25. Zhao, Z.; Zhang, L.; Bai, P.; Du, W.; Wang, S.; Xu, X.; Dong, Q.; Li, Y.; Han, B. Tribological Behavior of In Situ TiC/Graphene/Graphite/Ti6Al4V Matrix Composite Through Laser Cladding. *Acta Met. Sin. Engl. Lett.* **2021**, *34*, 1317–1330. [CrossRef]
26. Suresha, B.; Seetharamu, S.; Kumaran, P.S. Investigations on the influence of graphite filler on dry sliding wear and abrasive wear behaviour of carbon fabric reinforced epoxy composites. *Wear* **2009**, *267*, 1405–1414. [CrossRef]
27. Cui, Y.; Shen, J.; Manladan, S.M.; Geng, K.; Hu, S. Wear resistance of FeCoCrNiMnAlx high-entropy alloy coatings at high temperature. *Appl. Surf. Sci.* **2020**, *512*, 145736.1–145736.14. [CrossRef]
28. JWilson, J.E.; Stott, F.H.; Wood, G.C. The development of wear-protective oxides and their influence on sliding friction. *Proc. R. Soc. Lond. Ser. A Math. Phys. Sci.* **1980**, *369*, 557–574. [CrossRef]
29. Quinn, T. The oxidational wear of low alloy steels. *Tribol. Int.* **2002**, *35*, 691–715. [CrossRef]
30. Quinn, T.F.J. The Effect of "Hot-Spot" Temperatures on the Unlubricated Wear of Steel. *ASLE Trans.* **1967**, *10*, 158–168. [CrossRef]
31. Mishina, H.; Hase, A. Wear equation for adhesive wear established through elementary process of wear. *Wear* **2013**, *308*, 186–192. [CrossRef]
32. Wang, Y.B.; Li, L.; An, J. Dry wear behavior and mild-to-severe wear transition in an Mg-Gd-Y-Zr alloy. *Surf. Topogr. Metrol. Prop.* **2021**, *9*, 25032. [CrossRef]
33. Zheng, G.; Luo, X.; Yang, Y.; Kou, Z.; Huang, B.; Zhang, Y.; Zhang, W. The gradient structure in the surface layer of an Al-Zn-Mg-Cu alloy subjected to sliding friction treatment. *Results Phys.* **2019**, *13*, 102318. [CrossRef]
34. Xin, L.; Yang, B.; Wang, Z.; Li, J.; Lu, Y.; Shoji, T. Microstructural evolution of subsurface on Inconel 690TT alloy subjected to fretting wear at elevated temperature. *Mater. Des.* **2016**, *104*, 152–161. [CrossRef]
35. Zhang, W.; Lu, J.; Huo, W.; Zhang, Y.; Wei, Q. Microstructural evolution of AZ31 magnesium alloy subjected to sliding friction treatment. *Philos. Mag.* **2018**, *98*, 1576–1593. [CrossRef]
36. Machado, M.; Moreira, P.; Flores, P.; Lankarani, H.M. Compliant contact force models in multibody dynamics: Evolution of the Hertz contact theory. *Mech. Mach. Theory* **2012**, *53*, 99–121. [CrossRef]
37. Greiner, C.; Gagel, J.; Gumbsch, P. Solids Under Extreme Shear: Friction-Mediated Subsurface Structural Transformations. *Adv. Mater.* **2019**, *31*, e1806705. [CrossRef] [PubMed]
38. Alidokht, S.; Abdollah-Zadeh, A.; Assadi, H. Effect of applied load on the dry sliding wear behaviour and the subsurface deformation on hybrid metal matrix composite. *Wear* **2013**, *305*, 291–298. [CrossRef]
39. Mandal, M.; Mitra, R. Study of Dry Sliding Wear Behavior of Hot-Rolled and Mushy-State Rolled Al-4.5Cu-5TiB2 In-Situ Composite with Analysis of Work Hardening and Subsurface Microstructure-Microtexture Evolution Using EBSD. *Met. Mater. Trans. A* **2019**, *50*, 5356–5372. [CrossRef]
40. Hariharasakthisudhan, P.; Jose, S.; Manisekar, K. Dry sliding wear behaviour of single and dual ceramic reinforcements premixed with Al powder in AA6061 matrix. *J. Mater. Res. Technol.* **2019**, *8*, 275–283. [CrossRef]
41. Chen, X.; Han, Z.; Lu, K. Wear mechanism transition dominated by subsurface recrystallization structure in Cu–Al alloys. *Wear* **2014**, *320*, 41–50. [CrossRef]

crystals

MDPI

Article

Plastic Damage Assessment in 316 Austenitic Steel Using the Misorientation Parameters from an In Situ EBSD Technique

Xiao Wang [1,†], Zechen Du [2,†], Fubao Zhang [1], Yu Zhu [1], Yu Liu [1,*] and Hui Wang [3,*]

1 School of Mechanical Engineering, Nantong University, Nantong 226019, China
2 School of Civil and Resource Engineering, University of Science and Technology, Beijing 100083, China
3 Institute for Industrial Science, The University of Tokyo, 5-1-5 Kashiwanoha, Kashiwa 277-8574, Japan
* Correspondence: liuyu_me@ntu.edu.cn (Y.L.); hui_wang@metall.t.u-tokyo.ac.jp (H.W.)
† These authors contributed equally to this work.

Abstract: Plastic damage assessment in 316 austenitic steel was performed in this research by using the misorientation parameters derived from an in situ EBSD technique. With the increase in plastic strain, the misorientation parameters, such as the Grain Reference Orientation Deviation (GROD), Grain Orientation Spread (GOS), the Grain Orientation Spread over the grain Diameter (GOS/D), and Geometrically Necessary Dislocation (GND) density presented a growing trend. Nevertheless, the variation in GROD did not show a monotony trend, and the relative increase in the amplitude of GOS and GND density was less in the late plastic stage. Compared with the above parameters, the (GOS)/D exhibited a near-linear increase during the plastic tensile stage. As the specimen was stretched to a strain of 56.99%, the (GOS)/D increased 8.9 times compared with the original specimen. The results showed that the (GOS)/D parameter has the potential of becoming an indicator for the assessment of plastic damage in 316 steel.

Keywords: plastic damage; assessment; 316 steel; misorientation parameters

1. Introduction

Plastic strain-induced damage is one of the main failure modes in engineering components. Generally, the assessment of plastic damage in components is of great significance for the safety and reliability of engineering structures. The more accurate the diagnosis of component damage, the better the chances that treatment will work [1]. Nevertheless, accurately knowing the degree of material damage is not often that simple. Most of the time, the assessment methods based on hardness analysis, elongation surveying, or cavity quantification, are not precise, and seem increasingly inadequate for the higher demands of engineering protection.

The misorientation parameters measuring method based on the Electron backscatter diffraction (EBSD) technique has seen rapid development in recent years [2,3]. Since EBSD has excellent spatial resolution levels (~50 nm) and comes complete with sophisticated post-processing algorithms, it has become possible to make unprecedented measurements of the local lattice orientation at a sub-micron level. To date, some researchers have suggested that there is a close correlation between the lattice misorientation parameters and the plastic strain of a material. For instance, Kamaya et al. [4] showed that after a 14.4% plastic deformation, the local change in the crystal orientation was significant inside the grains of 304 austenitic steel; Wang et al. [5] indicated that with the increase in plastic strain, the Kernel average misorientation (KAM) obviously increased in mild ferrite/pearlite steel; Allain-Bonasso et al. [6] illustrated that, in a ferritic single-phase interstitial-free (IF) steel, the misorientation parameters such as Grain Orientation Spread (GOS) and KAM strongly increased after tensile deformation tests. However, the above research results were based on ex situ EBSD measurements (i.e., EBSD mapping was not made on the same sample).

Therefore, they failed to investigate the misorientation parameter variation in grains for the same piece of material during plastic deformation.

In this research, the in situ EBSD technique was utilized to overcome the limitations mentioned above. A miniature tensile stage was inserted into the specimen chamber of the SEM, so that the observation area of the sample was locked down, and the significant information could be extracted from the same area. Furthermore, a widely used austenitic steel, i.e., 316 stainless steel, was used here as the research object. It is hoped that the study can provide valuable insights into plastic damage assessment through the use of the misorientation parameters method.

2. Experiment

The chemical composition of the 316 stainless steel used in this study was: C: 0.04, Cr: 16.38, Ni: 9.73, Mo: 2.15, Mn: 0.932, Si:0.492, Cu: 0.275, Co: 0.351, and Fe: Bal (wt. %). In order to conduct the in situ EBSD experiment, the 316 steel specimen was fabricated from the hot-rolled sheet and then processed into the dimensions shown in Figure 1a. The Zeiss-Sigma 300 Scanning Electron Microscope (SEM) and Oxford-Symmetry B EBSD detector were used. The microscope was operated at 20 kV. The EBSD scanning area was 300.0 μm × 270.0 μm and the step size was 300 nm. The type of tensile stage was Kammrath & Weiss (KW) (see Figure 1b). The loading rate of the 316 stainless steel specimen was 10 μm/s. The stress–strain diagram is displayed in Figure 1c. The fracture-strain of this 316 -steel specimen was 68.76%. Specific stress–strain values are shown in Table 1. The specimen geometry was described by the loading direction (LD), transverse direction (TD) and normal direction (ND). The EBSD measurements were conducted on the external surface (A), at the center of tensile specimen (see Figure 1a). The EBSD measurement results were analyzed using ATEX software [7]. It is worth noting that the LD was parallel to the RD of the plate (see the sampling diagram in Figure 1a).

Table 1. Stress–strain data of the 316 stainless steel corresponds to each stretch point.

Stretch Point	1	2	3	4	5	6	7
Strain (%)	6.27	12.58	19.43	25.71	31.37	37.33	45.86
Stress (Mpa)	264.92	302.13	330.10	341.87	352.16	364.79	378.47

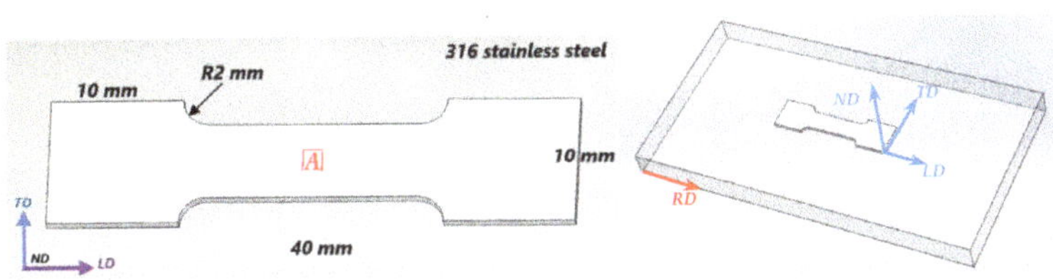

(a)

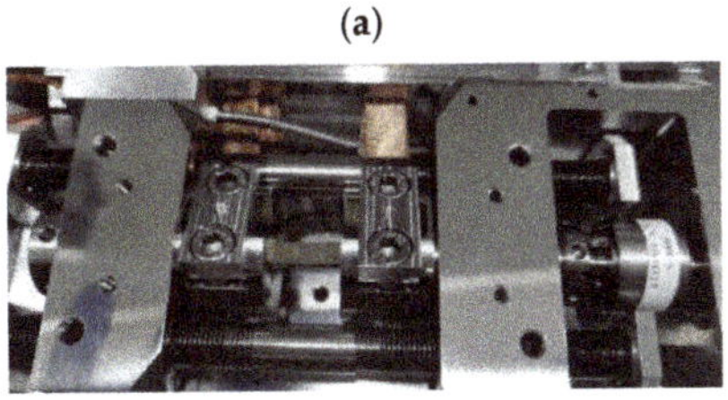

(b)

Figure 1. *Cont.*

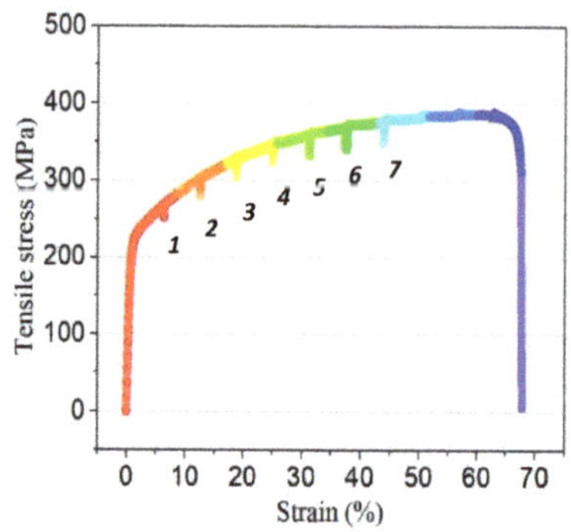

(c)

Figure 1. The in situ tensile experiment of 316 steel specimen. (**a**) The specimen size diagram; (**b**) The KW tensile stage; (**c**) The engineering stress–strain diagram.

3. Results and Discussion

The EBSD results of 316 steel at different plastic stages are shown in Figure 2. The measurement covered more than four hundred grains. The grain statistical analyses were conducted using an ATEX built-in procedure. The grain reference orientation deviation (GROD) maps were used first to characterize the plastic damage of the 316 steel specimen (see Figure 2a). The GROD is the misorientation between the orientation of each pixel of the grain and the average orientation of the grain, which can be expressed as: $GROD_i = |o_i - o_j|$, where o_i is the orientation of a pixel point i, and o_j is the orientation of its neighbor point, j. From Figure 2a, it can be observed that the GROD value is low in the original specimen (inside the grains), but significantly increased as the specimen was stretched to the plastic stage (i.e., Point 1, stretched stress: 264.92 Mpa, strain: 6.27%). Furthermore, it can also be recognized from GROD maps that the GROD values are higher close to the grain boundaries (see white arrow), while they are relatively low in the center of the grain (see black arrow). The result indicates that in 316 steel, the largest deformation and distortion in grain preferentially occurs at the region near the grain boundaries, especially with grains of a larger dimension.

The Grain Orientation Spread (GOS) maps are shown in Figure 2b. The GOS was the average misorientation between the orientations of each pixel and the average orientation of grain, per grain. From Figure 2b, it can be observed that the GOS value in 316 steel progressively increased with the increase in plastic strain, while the image quality decreased during this process, manifested by the increasing number of pixel points that could not be indexed. Previously, Saeidi et al. [8] investigated the GOS distribution in a deformed, dual-phase (Ferrite-Martensite) steel, and suggested that the GOS value was higher in the larger grains compared to the smaller grains, and that the GOS could be used as an indication of strain distribution. However, the results here do not completely agree with those findings. For instance, two grains of 'A' (size: 4.20 um, GOS: 0.0237°) and 'B' (size: 2.15 um, GOS: 0.0129°) were chosen in the original specimen, and it can be observed that after the plastic deformation (i.e., Point 2, Stretched stress: 302.13 Mpa, Strain: 12.58%), it is the smaller grain 'B' that has a higher GOS value. This result indicates that the grain size was not the only essential factor that affects the GOS level; apart from the GOS parameter, Allain-Bonasso et al. [6] revealed that the (GOS)/D parameter was also sensitive to the plastic deformation of grain populations in an IF steel. The (GOS)/D was the Grain Orientation Spread over the grain Diameter (D), which can be utilized here to study the deformation degree of 316 steel. For this research, the gradual increase in (GOS)/D values in 316 steel can be directly observed in the (GOS)/D maps, as shown in Figure 2c. The subpopulation of the smallest grains showed the highest GOS/D values, which is consistent with the results in [6].

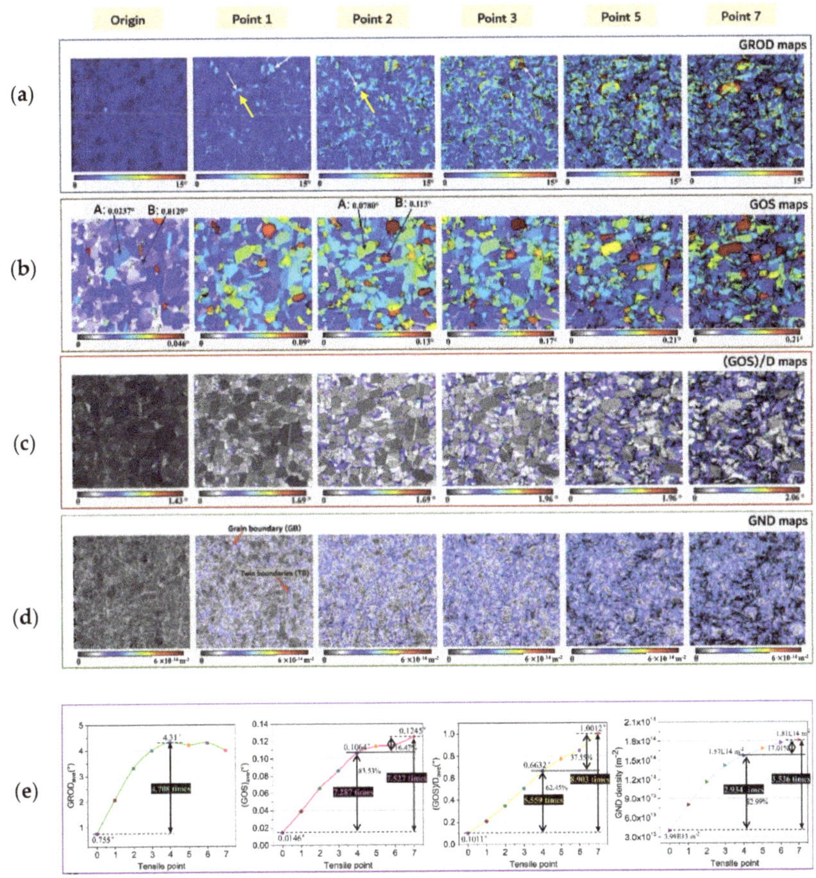

Figure 2. The EBSD measurements of the 316 steel with different plastic strain. (**a**) GROD maps; (**b**) GOS maps; (**c**) (GOS)/D maps; (**d**) GND density maps; (**e**) The variation of above parameters in 316 steel during the tensile process.

In addition to the misorientation parameters shown above, the geometrically necessary dislocation (GND) density parameter was also used to investigate plastic damage in 316 steel. The GND density can be calculated from the misorientation data between the neighbor points [9], which represents the distortion of the crystal lattice introduced by the plastic deformation. Figure 2d shows the variation in GND density in 316 steel in the plastic tensile stage. The continuous increase in the GND density of 316 steel can be seen as the plastic deformation increases. Meanwhile, it should be noted that, inside the grains, the distribution of GND density is inhomogeneous, and tended to centralize in the vicinity of the grain boundaries and the twin boundaries. Furthermore, it should be also pointed out that no evidence of correlation can be found between the level of GND value and grain size in 316 steel.

From Figure 2a–d, it can be recognized that, with the increase in plastic strain, the misorientation parameters, such as the GROD, GOS, (GOS)/D, and GND density, all presented a growing trend, which means that these parameters can be used as indicators for the quantitative assessment of plastic damage. Therefore, it is necessary to understand how assessable these parameters are. Since the data value of each pixel can be extracted from the misorientation parameter maps, the average processing was also performed here to calculate the average value of each map, as shown in Figure 2a–d.

Figure 2e shows the variation in the above parameters in 316 steel during the tensile stage. The tensile point 0 is the specimen in the original state. The $GROD_{ave}$ presented a clear upward trend in the early-to-mid-plastic stage, i.e., from Point 0 to Point 4 (Stretched stress: 341.87 Mpa, Strain: 25.71%), which increased more than 4.7 times. However, after this, the $GROD_{ave}$ did not show regularity with the increase in plastic strain; the $GROD_{ave}$ fluctuated within the range from Point 4 to Point 7 (Stretched stress: 302.13~378.47 Mpa, Strain: 25.71~45.86%), and then decreased in the last stage. Furthermore, it can be found that the rapid rise in $GROD_{ave}$ occurs just in the range of Point 0 to Point 2 (Stretched stress: 0~302.13 Mpa, Strain: 0~12.58%), since its growth slowed down as the specimen was stretched to Point 3 (Strain: 19.43%). Hence, it can be inferred that the $GROD_{ave}$ may only be suitable for the early assessment of the degree of degradation in 316 steel.

Different from $GROD_{ave}$, the GOS_{ave} and $(GOS)/D_{ave}$ monotonously increased with the increase in plastic strain. Nevertheless, there is also a visible difference in the variation rules of both parameters. From Figure 2e, it can be observed that the GOS_{ave} sharply increased in the early-to-mid-plastic stage, and from Point 0 to Point 4, the increase in GOS_{ave} rose 7.2 times, which accounted for more than 83.5 percent of the total amount. However, in the subsequent tensile steps, the growth rate of GOS_{ave} dropped significantly, especially in the range from Point 5 to Point 6. Thus, the GOS_{ave} parameter may not be very well-suited to the assessment of plastic damage in 316 stainless steel in the later stage; unlike GOS_{ave}, it seems that the $(GOS)/D_{ave}$ exhibited a near-linear increase with the increase in tensile strain. Although the extent of the $(GOS)/D_{ave}$ increase in the early-to-mid-plastic stage (i.e., From Point 0 to Point 4) was less than the GOS_{ave} (i.e., 5.559 < 7.287), the extent of the increase in the former in the subsequent plastic stages was still evident, which accounted for a higher proportion of the total when compared with the GOS_{ave} (37.55% > 16.47%). Furthermore, from Point 0 to Point 7, the $(GOS)/D_{ave}$ rose 8.9 times, which was already much larger than the GOS_{ave} increase (i.e., 7.527 times). Thus, compared with the GOS_{ave}, the $(GOS)/D_{ave}$ parameter may be a better option for the assessment of plastic damage in 316 steel.

Apart from the above parameters, a progressive increase in GND_{ave} was also observed in 316 steel during the plastic tensile stage. However, the increasing amplitudes of GND_{ave} in the range of Point 0–4 and Point 0–7 were just 2.934 times and 3.536 times, respectively, which were much less than the GOS_{ave} and (GOS)/Dave. Furthermore, the extent of the increase in GND_{ave} in the later plastic stage (i.e., From Point 4 to Point 7), which just accounted for 17.01% of the total increment, was also not evident. Therefore, the GND_{ave} parameter may be not the optimal choice here.

The orientation variation in 316 steel during the plastic tensile stage can also be recognized in terms of the macroscopic viewpoint in the development of crystallographic texture. Pole figure (PF) can be used to provide a classical representation of orientation. The PF is a two-dimensional graphical representation of orientation which displayed the orientation of the selected normal plane (i.e., a pole) with respect to the sample reference frame. Figure 3a shows the PFs of the original 316 steel specimen. When comparing textural intensities, it can be recognized that the (111) and (110) textures have relatively strong basal textures, with a maximum intensity of 3.74 mud and 3.64 mud, respectively. This indicates that more lattice planes are aligned in these two directions than in a sample with random texture; Figure 3b show the (111) and (110) textures of 316 steel grains at the different plastic stages. It can be seen that, after the plastic tensile stage, the intensity of both textures changed noticeably. The maximum intensity in the (111) texture of the Point 1-stretched specimen was 3.77 mud, while it increased to 3.97 mud as the specimen was stretched to Point 7. Conversely, a strong texture with an intensity of 3.68 mud was observed in the (110) texture of the Point-1-stretched specimen, but it decreased to 3.37 mud at Point 7. From the above results, the grain rotation of 316 steel during the tensile process can be clearly recognized. The grain deformation led to the variation in misorientation parameters, and thus the plastic damage can be quantitatively assessed using the misorientation parameters method.

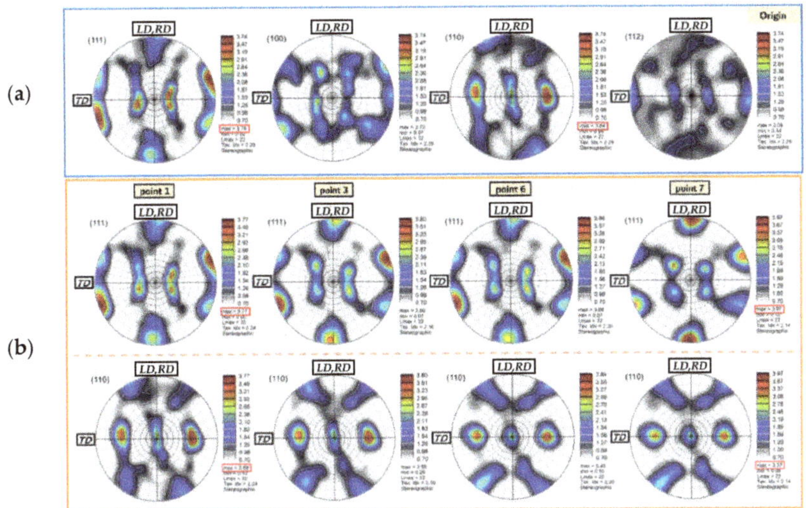

Figure 3. The pole figures of the 316 steel before and after the plastic deformation. (**a**) The (111), (100), (110), (112) pole figures of the original 316 steel; (**b**) The (111) and (110) pole figures of the 316 steel at the different plastic stages.

4. Conclusions

The assessment of plastic damage in a 316 austenitic stainless steel was studied using the misorientation parameters from an in situ EBSD measurement. The results demonstrated that with the increase in plastic strain, the misorientation parameters, such as the GROD, GOS, (GOS)/D, and GND density, presented a growing trend. However, the extent of the variation in the above parameters was different. Among them, the (GOS)/D exhibited a near-linear increase. As the 316 steel was stretched from the original state to a strain of 56.99%, the (GOS)/D increased 8.9 times, which was a much larger increase than seen in the other parameters. Therefore, the (GOS)/D parameter has the potential to become an indicator for the assessment of plastic damage in 316 steel.

Author Contributions: Conceptualization, X.W.; validation, Z.D.; formal analysis, Y.Z.; resources, Y.L. and H.W.; data curation, F.Z.; writing—original draft preparation, X.W.; writing—review and editing, Z.D.; visualization, F.Z.; supervision, Y.L. and H.W.; funding acquisition, X.W. and H.W. All authors have read and agreed to the published version of the manuscript.

Funding: This research was funded by High level introduction of talent research start-up fund, Nantong university, grant number 2022078.

Data Availability Statement: Not applicable.

Acknowledgments: The authors would like to express their gratitude for projects supported by the high-level introduction of talent research start-up fund, Nantong university (Grant No. 2022078), and the Analysis and Test Centre of Nantong University.

Conflicts of Interest: No conflict of interest exits in the submission of this manuscript, and manuscript is approved by all authors for publication.

References

1. Wang, X.; Chen, J.G.; Su, G.F.; Cui, C.; Li, H.Y. Application of electromagnetism method to characterize the degradation behavior in structural mild steel within the elastic range. *Constr. Build. Mater.* **2020**, *118011*, 241. [CrossRef]
2. Yoda, R.; Yokomaku, T.; Tsuji, N. Plastic deformation and creep damage evaluations of type 316 austenitic stainless steels by EBSD. *Mater. Charact.* **2010**, *61*, 913–922. [CrossRef]
3. Fujiyama, K.; Mori, K.; Kaneko, D.; Kimachi, H.; Saito, T.; Ishii, R.; Hino, T. Creep damage assessment of 10Cr-1Mo-1W-VNbN steel forging through EBSD observation. *Int. J. Press. Vessel. Pip.* **2009**, *86*, 570–577. [CrossRef]
4. Kamaya, M. Assessment of local deformation using EBSD: Quantification of local damage at grain boundaries. *Mater. Charact.* **2012**, *66*, 56–67. [CrossRef]
5. Wang, X.; Chen, J.G.; Su, G.F.; Li, H.Y.; Wang, C. Plastic damage evolution in structural steel and its non-destructive evaluation. *J. Mater. Res. Technol.* **2020**, *9*, 1189–1199. [CrossRef]
6. Allain-Bonasso, N.; Wagner, F.; Berbenni, S.; Field, D.P. A study of the heterogeneity of plastic deformation in IF steel by EBSD. *Mater. Sci. Eng. A* **2012**, *548*, 56–63. [CrossRef]
7. Beausir, B.; Fundenberger, J.J. Analysis Tools for Electron and X-ray Diffraction, ATEX-Software, Université de Lorraine-Metz. 2017. Available online: http://www.atex-software.eu/ (accessed on 9 August 2022).
8. Saeidi, N.; Ashrafizadeh, F.; Niroumand, B.; Barlat, F. EBSD Study of Damage Mechanisms in a High-Strength Ferrite-Martensite Dual-Phase Steel. *J. Mater. Eng. Perform.* **2015**, *24*, 53–58. [CrossRef]
9. Zhou, X.W.Z.Q.; Liu, S.; Huang, M.Y. Investigation of the evolution of Geometrically Necessary Dislocation (GND) tensor in a type 316 steel by using In-situ EBSD technique. *Mater. Lett.* **2021**, *286*, 129254.

crystals MDPI

Article

Nanoindentation Stress Relaxation to Quantify Dislocation Velocity–Stress Exponent

Tzu-Yi Chang [1], Gavin Vandenbroeder [1], David M. Frazer [2,†], Dewen Yushu [2], Stephanie Pitts [2] and Tianyi Chen [1,*]

[1] School of Nuclear Science and Engineering, Oregon State University, Corvallis, OR 97330, USA; changtz@oregonstate.edu (T.-Y.C.); vandenbg@oregonstate.edu (G.V.)

[2] Idaho National Laboratory, 1955 Fremont Ave, Idaho Falls, ID 83415, USA; david.frazer@ga.com (D.M.F.); dewen.yushu@inl.gov (D.Y.); stephanie.pitts@inl.gov (S.P.)

* Correspondence: tianyi.chen@oregonstate.edu; Tel.: +1-(541)737-8190

† Current address: General Atomics, San Diego, CA 92121, USA.

Abstract: This work reports a new methodology using indentation stress relaxation to characterize the dislocation velocity–stress exponent. Through the indentation stress relaxation process, the dislocation structure builds up at the rate governed by dislocation velocity, which is a function of the externally applied stress. The relationship between the dislocation velocity and stress can thus be derived from the indentation stress relaxation data of the stress as a function of time. In this study, instrumented nanoindentation stress relaxation experiments were performed on pure aluminum samples, following three different initial displacement rates of 100, 400, and 800 nm/s. Based on the scaling properties of dislocation kinetics, the data were interpreted to derive a dislocation velocity–stress exponent of 2.5 ± 0.5 for room-temperature aluminum. Crystal plasticity finite-element simulations were performed to illustrate the sensitivity of the proposed nanoindentation stress relaxation methodology to the dislocation velocity–stress exponent value.

Keywords: scaling; nanomechanics; creep; activation volume

1. Introduction

Emerging instrumented nanoindentation techniques are facilitating the high-throughput characterization of local mechanical properties, especially in materials with microstructural heterogeneity and limited available volume, such as in nuclear materials, thin films, composites, and functionally graded materials [1–5]. However, the gradients of stress and strain under indentation create barriers to data interpretation. Strategies utilizing the nature of the stress fields of different indentation tip geometries have been applied to interpretate the nanoindentation data. For instance, the self-similarity of a Berkovich indenter facilitates the extrapolation of bulk hardness from indentation data subjected to the size effect [6], as well as the determination of the strain rate–stress exponent, n, of the strain rate from nanoindentation creep experiments [7–10]. However, as pointed out by Goodall and Clyne, the strain rate–stress exponent determined by the nanoindentation creep methodology is rarely consistent with that determined by conventional bulk creep testing [10]. Another example is the utilization of a flat-punch indenter to mitigate the stress gradient, as seen in several indentation creep and stress relaxation studies [11,12]. Yet, with the reduction in indenter size, the heterogeneity of the stress field at the periphery of a flat punch becomes more significant; the existing data interpretation model of the flat-punch stress relaxation also relies on the presumption of dislocation density invariance [11]. Utilizing established scaling relationships among the stress, dislocation kinetics, and time [13], we present a new approach to determine the dislocation velocity–stress exponent, m, through the nanoindentation stress relaxation method with a Berkovich indenter. In the following section, we introduce the basis of dislocation kinetics concerning the importance of m, the general stress relaxation process associated with the coevolution of stress components and dislocations. We also discuss the theorical basis for the development of a new nanoindentation stress relaxation methodology.

Plastic strain in crystalline materials is typically mediated by dislocations. Under applied stress, dislocation generation, motion, interaction, and annihilation govern both the macroscopic plastic straining and the flow stress. Notably, the relationship between dislocation velocity and stress is a key to the rate of plastic deformation. For example, it influences the kinetics of dislocation interactions that form networks, the strain rate under steady-state dislocation density, and the plastic energy dissipation during ductile fracture. This relationship has been described by a power law, shown in Equation (1), which is commonly used in both reduced-order models and physics-based models, including the crystal plasticity (CP) finite element method and dislocation dynamics simulations [14].

$$v = B(\sigma - \sigma_i)^m, \tag{1}$$

In Equation (1), v is the dislocation velocity, B is a material constant depending on the microstructure profile, σ is the applied stress, σ_i is the internal stress, and m is a velocity–stress exponent [15,16]. m relates to the mechanism of dislocation mobility. For example, $m \approx 1$ corresponds to the viscous motion of dislocations, e.g., when a solute cloud is formed and exerts a drag force on the dislocation. A much larger m ($\gg 1$) corresponds to the jerky movement of dislocations as the pinned dislocations break away from obstacles [16]. The unpinning process may involve thermal activation, including diffusion, for dislocations to overcome the obstacles and move freely before encountering the next obstacle.

The dislocation velocity–stress relationship and its macroscopic manifestation, the strain rate–stress relationship, can be investigated using stress relaxation experiments [11,17–20]. Following transient loading, the total strain is held constant, allowing the elastic strain, ε_e, to convert to plastic strain, ε_p, over time. As a result, the rate of stress relaxation, $\dot{\sigma}$, is controlled by the plastic strain rate, $\dot{\varepsilon_p}$:

$$\dot{\sigma} = E\dot{\varepsilon_e} = -E\dot{\varepsilon_p}, \tag{2}$$

where E is the Young's modulus of the material. The rate of plastic strain accumulation relates to the dislocation mobility via the Orowan equation [14]:

$$\dot{\varepsilon}_p = \frac{1}{M} b\rho v, \tag{3}$$

where b is the Burgers vector, ρ is the dislocation density, and M is the Taylor factor homogenizing the dislocation motion in different grain orientations. Equation (3) assumes a constant ρ and σ_i through the stress relaxation transient, so the dislocation velocity–stress exponent can be derived from the experimentally observed $\sigma - t$ curve, where t stands for the time passed since the start of the strain holding period [17,19]. However, the plastic straining processes are often associated with dislocation accumulation and evolution, supposedly changing both ρ and σ_i. As reported in [17,21], the assumptions of constant ρ and σ_i, leading to an inaccurate estimation of the effect stress, $\sigma*$, have led to errors in the estimated stress components as large as 100% [21].

In this work, we introduce an indentation-based stress relaxation methodology, whose data interpretation does not require the assumption of a fixed dislocation density. Instead, ρ and σ_i are interdependent, based on their scaling relationship. For simplification and coherence with our selected model alloy system, aluminum, we employ the Orowan hardening model for this relationship [16,22]:

$$\sigma_i = \alpha M b G \sqrt{\rho}, \tag{4}$$

where α is the dislocation interaction constant and G is the elastic shear modulus. We also consider that the rate of dislocation accumulation correlates to plastic straining:

$$\dot{\rho} = \frac{M}{b}\frac{1}{\Lambda}\dot{\varepsilon}_p, \tag{5}$$

where Λ is the mean free path (MFP) of dislocations and approximates to $1/\sqrt{\rho}$, with the absence of precipitates or grain boundaries, such as in pure metals under the nanoindentation condition. The above equations lead to the following:

$$\dot{\sigma} = -\frac{Eb}{M}\frac{\sigma_i^2}{(\alpha MbG)^2}B(\sigma - \sigma_i)^m, \tag{6}$$

$$\dot{\sigma}_i = \frac{\alpha MG}{2b}\frac{\sigma_i^2}{(\alpha MbG)^2}B(\sigma - \sigma_i)^m, \tag{7}$$

Defining the applied stress at the beginning of the stress relaxation process to be σ_0, the evolutions of σ and σ_i can be described by the dimensionless parameters σ^+ and σ_i^+:

$$\sigma = \sigma^+\sigma_0, \tag{8}$$

$$\sigma_i = \sigma_i^+\sigma_0, \tag{9}$$

σ^+ and σ_i^+ have initial values of 1 and 0, respectively, and they decrease and increase, respectively, until $\sigma^+ = \sigma_i^+$. Thus, Equation (6) can be written in the following form:

$$\dot{\sigma} = -\frac{BEb}{M(\alpha MbG)^2}\sigma_0^{m+2}\left(\sigma_i^+\right)^2\left(\sigma^+ - \sigma_i^+\right)^m, \tag{10}$$

In Equation (10), $\dot{\sigma}$, is kept dimensional to be consistent with the experimental measurement in this work. The stress relaxation rate scales with the initial stress via the following:

$$\dot{\sigma} \sim \sigma_0^{m+2}, \tag{11}$$

The stress gradient forms underneath the Berkovich indenter, during the nanoindentation. The deeper a point is from the contact surface, the lower the stress that point experiences. Via Equations (1) and (3), the kinetics of stress relaxation depend on the local stress and thus on the depth from the contact surface. The stress relaxation process is faster at a shallower point, closer to the contact surface, than at a deeper point, which is further away. We assign d_{eff} and σ_{eff} as the effective depth and stress of the region, respectively, where stress relaxation primarily occurs at the beginning of the indenter-holding phase. Then, σ_{eff} scales inversely with the time elapsed during the indenter loading, t_{load}, as d_{eff} increases with t_{load}, during which time the stress relaxation has already taken place in the shallower region where the kinetics are faster. Defining the indenter pre-loading rate to be $\dot{h}$ ($\dot{h} \sim 1/t_{load}$), the scaling relationship between σ_{eff} and $\dot{h}$ can be derived based on the theorems of dislocation similitude in [13]:

$$\sigma_{eff} \sim t_{load}^{\left(-\frac{1}{m+1}\right)} \sim \dot{h}^{\left(\frac{1}{m+1}\right)}, \tag{12}$$

where the indenter loading rate $\dot{h}$ is inversely proportional to t_{load}. Recognizing that this σ_{eff} is the initial stress σ_0 during the stress relaxation in Equation (10), we obtain the scaling relationship between the simultaneous stress relaxation rate, $\dot{\sigma}$, and the indenter displacement rate before the relaxation, $\dot{h}$:

$$\dot{\sigma} \sim \sigma_{eff}^{m+2} \sim \dot{h}^{\left(\frac{m+2}{m+1}\right)}, \tag{13}$$

With this scaling relationship, we designed and performed the following experiments with varying $\dot{h}$ and derived the dislocation velocity–stress exponent m using the obtained $\dot{\sigma} \sim \dot{h}$ relationship. We further employed crystal plasticity (CP) simulations to illustrate the abovementioned temporal and spatial behavior of nanoindentation stress relaxation. The CP simulations showed the depth dependence of stress relaxation kinetics on $\dot{h}$, which underpins Equation (13).

2. Materials and Methods

Commercial pure aluminum (99.9% purity) was used in this study. A specimen of $10 \times 8 \times 6$ mm^3 was mechanically ground with silicon carbide sandpapers of 600, 800, and 1200 grits, polished with 0.1 μm alumina suspensions, and vibratory polished with 0.05 μm alumina suspensions. This creates a mirror-like flat surface with minimal residual stress for nanoindentation experiments.

We used instrumented nanoindentation (FT-NMT04, FemtoTool, Buchs, Zwitserland) with a diamond Berkovich tip to perform the stress relaxation testing at room temperature. The indentation was first loaded in displacement-controlled mode with various displacement rates of 100, 400, and 800 nm/s. As soon as the maximum depth of 800 nm was reached, the tip was held at this depth for 60 s, during which the applied load was monitored. For each displacement rate, the dataset comprising 10 indents was analyzed to provide statistical reliability. The interspacing of the indents was greater than 20 μm.

Crystal plasticity finite element method simulations of the stress relaxation process were performed using Idaho National Laboratory's Multiphysics Object-Oriented Simulation Environment (MOOSE 2.0) software [23]. MOOSE uses a continuous update system, with revision history tracked by *git*. These simulations used the version of MOOSE current in 02_2023. The Kalidindi model was adopted as reported in [24]. The model parameters are provided in Table 1.

Table 1. Parameterization for aluminum CP model.

$\dot{\gamma}_0$	Reference Slip Rate	0.001 s^{-1}
h_0	Hardening Constants	75 MPa
g_0	Initial Slip Resistance	31 MPa
g_{sat}	Saturated Slip Resistance	63 MPa
xm	Strain Rate Sensitivity	0.333
q	Latent Hardening Constant	1.4
a	Hardening Exponent	2.25
C_{11}	Elasticity Tensor	107 MPa
C_{12}	Elasticity Tensor	61 MPa
C_{44}	Elasticity Tensor	28 MPa

The simulation used a Berkovich tip with semi-angle of 65.27°, an effective cone angle of 70.3°, and a roundness of 20 nm [25]. The tip was inserted in the <001> direction of the face center cubic crystal, simulating aluminum. The displacement control was conducted in the CPFE simulation with various loading rates of 100, 800, and 1600 nm/s. The tip was held at a depth of 400 nm for 15 s to simulate the stress relaxation process. As shown later, a holding time of 15 s is sufficient for the stress relaxation of interest to fully take place. The indentation depth was reduced to 400 nm from the experimental value of 800 nm due to the limitation of computational power. Similarly, only one indentation orientation of <001> was selected to represent the general cases. However, such modification and simplifications should not qualitatively change the relationship between the stress relaxation process and the pre-loading rate, which is the key to the computational investigation in this work.

3. Results

Figure 1a presents examples of the experimental stress relaxation data following different initial loading rates. In all the cases, the stress-reduction rate is rapid at the beginning and gradually slows down towards the end of stress relaxation. The transition from the rapid/transient relaxation to the slower phase can be seen in Figure 1b, zooming into the initial relaxation period in Figure 1a. The slope of the initial stress relaxation correlates to the pre-loading rate, as expected from the above scaling analysis.

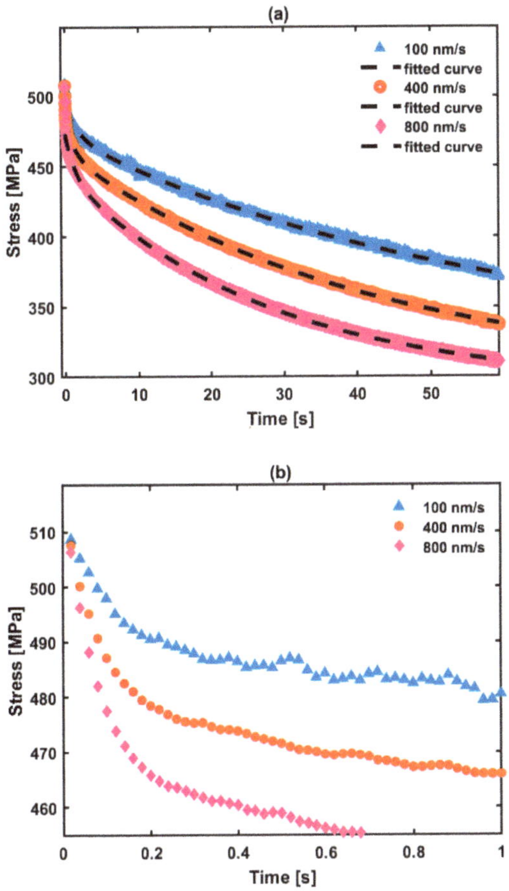

Figure 1. Nanoindentation stress relaxation curves following pre-loading at various displacement rates (**a**) and the initial transient relaxation within the first second (**b**).

The evolution of stress with time can be described by the sum of multiple exponential terms:

$$\sigma(t) = \Delta\sigma_1 e^{(-X_1 t_1)} + \Delta\sigma_2 e^{(-X_2 t_2)} + \cdots + \Delta\sigma_n e^{(-X_n t_n)} + \sigma_{sat}, \quad (14)$$

where $n = 1, 2, 3, \ldots$ indicates different rate components of stress relaxation, due to different relaxation mechanisms and various microstructures [26,27]. $\Delta\sigma_n$ is the stress drop relative to initial stress, and X_n is a stress decay constant with an inverse time unit. The stress relaxation data in this study fit to a double exponent expression, Equation (15). The fitting parameters are presented in Table 2.

$$\sigma(t) = \Delta\sigma_1 e^{(-X_1 t_1)} + \Delta\sigma_2 e^{(-X_2 t_2)} + \sigma_{sat}, \quad (15)$$

Table 2. Fitting parameters of double exponent at various loading rates.

Pre-Loading Rate (nm/s)	800	400	100
$\Delta\sigma_1$ (MPa)	61 ± 2	44 ± 4	23 ± 1
X_1 (/s)	3.2 ± 0.4	1.9 ± 0.5	0.67 ± 0.01
$\Delta\sigma_2$ (MPa)	150 ± 10	163 ± 7	154 ± 6
X_2 (/s)	0.037 ± 0.01	0.027 ± 0.004	0.018 ± 0.004
σ_{sat} (MPa)	300 ± 22	340 ± 20	363 ± 3

$\Delta\sigma_1$ and X_1 correspond to the rapid/transient relaxation within the first ~0.1 s, shown in Figure 1b, whereas $\Delta\sigma_2$ and X_2 correspond to the following slower relaxation process. The kinetics of the rapid/transient relaxation are strongly dependent on the pre-loading rate, whereas the kinetics during the slow relaxation are less so. The loading-rate dependency of the initial stress relaxation agrees with the above theoretical analysis of Equation (13). That is, a higher pre-loading rate leads to a higher effective stress, causing more rapid kinetics of stress relaxation. On the other hand, the secondary relaxation, characterized by $\Delta\sigma_2$ and X_2, is believed to be due to the dynamics equilibrium process as dislocation generation is balanced by annihilation.

Therefore, the rate of stress change due to dislocation accumulation at the beginning of the stress relaxation can be estimated by the product of $\Delta\sigma_1$ and X_1 (i.e., $\dot{\sigma} = \Delta\sigma_1 \times X_1$). Equation (13), indicates that the stress-changing rate and pre-loading rate have an exponential relationship:

$$ln(\dot{\sigma}) \sim \frac{m+2}{m+1} ln\left(\dot{h}\right), \quad (16)$$

The $\ln(\dot{\sigma}) - \ln\left(\dot{h}\right)$ relation is shown in Figure 2; the slope was fitted to be 1.3 ± 0.3 within 95% confidence bounds. Therefore, the dislocation velocity—exponent m—is derived to be 2.5 ± 0.5, according to Equation (16).

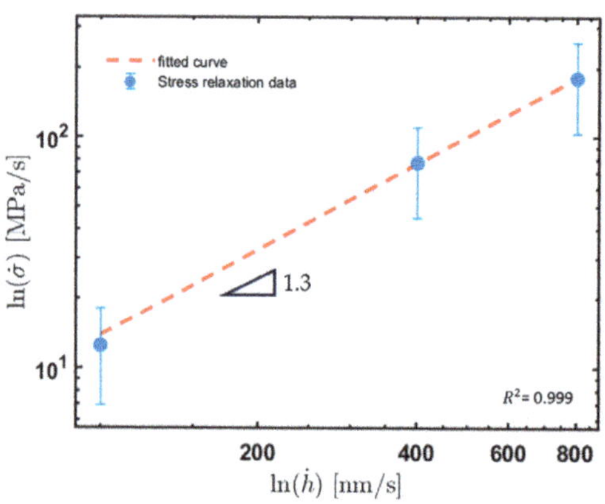

Figure 2. Correlation between transient stress-changing rate due to dislocation accumulation and pre-loading rate.

4. Discussion

4.1. Data Validation

This value of m indicates that the dislocation velocity is governed by both the thermal activation of dislocation climb and its velocity, dominated by the MFP of a free dislocation. The former depends on the diffusion coefficient and the latter on stress [28]. Considering that vacancies are mobile in aluminum at room temperature [29], and that the dislocation MFP will be limited by the high-density dislocation network formed under indentation [8], our derived value of m of 2.5 ± 0.5 appears reasonable.

Furthermore, based on Equations (3) and (4), one can derive the strain rate–stress exponent, n, of the plastic strain rate, $\dot{\varepsilon}_p$, as equal to $m + 2$. That result is because the dislocation density, ρ, and velocity, v, in Equation (3) have stress exponents of 2 and m, respectively [30]. Therefore, the strain rate–stress exponent can be derived using the reported nanoindentation stress relaxation method. In the room-temperature aluminum, we found the strain rate–stress exponent, n, to be 4.5 ± 0.5, which agrees with that derived from macroscopic creep testing, i.e., between 4 and 5 [10,31].

It is noteworthy that the strain rate–stress exponent determined by this new method aligns well with the values obtained from bulk creep testing. In contrast, the strain rate exponent derived from nanoindentation creep tests does not show this consistency. According to [10], nanoindentation creep tests are influenced by potential changes in dislocation density during deformation, as well as by significant gradients in stress, strain, and strain rate beneath the indentation. The proposed method, however, inherently accounts for these variables through the scaling relationships outlined in Equations (10)–(13).

4.2. Hypothesis Verification

As presented in the Introduction, the theory behind this reported method relies on a hypothesis that, immediately following the indentation loading, spontaneous stress relaxation occurs at a shallower region, with respect to the indenter tip, as the pre-loading displacement rate increases. In the following section, we discuss how this hypothesis is verified using both the experimentally derived dislocation activation volume, as a function of the pre-loading rate, and CP simulations to spatially resolve the stress relaxation process.

Activation volume (V) represents the volume of a dislocation involved in the rate process. It is fundamentally defined based on the dislocation length (l), Burgers vector (b), and its MFP (Δx) [32], as

$$V = lb\Delta x, \tag{17}$$

The activation volume under transient relaxation can be related to the stress drop over time [33]:

$$\Delta\sigma = -\frac{kT}{V} ln(1 + At), \tag{18}$$

where $\Delta\sigma$ is the stress drop, k is a Boltzmann constant, T is temperature, A is a time constant, and t is time. By fitting the stress relaxation curves in Figure 1 to Equation (18), we obtain the dislocation activation volumes under various pre-loading rates, as shown in Figure 3. Under a higher displacement rate, the smaller activation volume implies a smaller dislocation MFP. This reduction in the MFP is because the dislocation relaxation occurs at a shallower region where the dislocation density is higher. The uncertainties of the activation volumes are believed to have two sources. Firstly, the noise from the instrument becomes more significant at a lower pre-loading rate as the stress relaxation becomes less significant (Figure 1). Secondly, for each pre-loading rate, indentations are performed on multiple grains with different grain orientations, varying the dislocation network density underneath the indenter, as reported in [8].

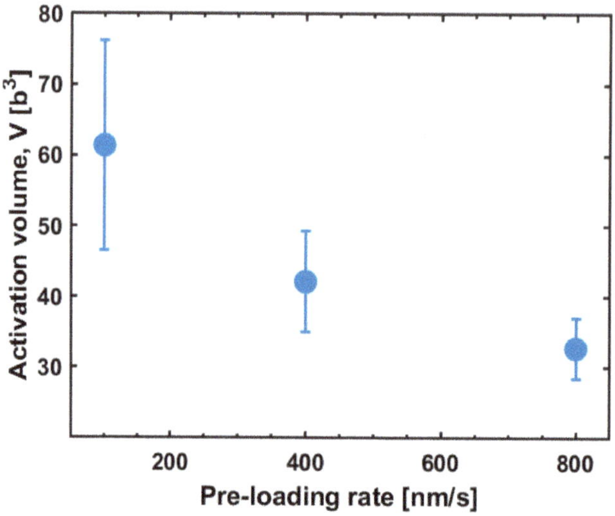

Figure 3. Activation volume of transient stress relaxation as a function of indentation pre-loading rate.

Figure 4 shows the snapshots of the CP simulations taken at 0, 1, and 15 s, following the pre-loading in Figure 4a, Figure 4b, and Figure 4c, respectively. The von Misses stress distributions in samples with pre-loading rates of 100 and 800 nm/s are shown in the left and right panels, respectively. As shown in Figure 4a, the higher initial stress in the faster-loaded sample, on the right, indicates that less stress relaxation had happened during the loading period, supporting the abovementioned hypothesis. During the first second, the stress relaxation is more significant in the faster-loaded sample, agreeing with the trend observed in Figure 1 and the $\Delta\sigma_1$ and X_1 values in Table 1. At 15 s, similar stress states are reached between the two samples, agreeing with the fact that the steady-state relaxation is less sensitive to the pre-loading rate, as shown by $\Delta\sigma_2$ and X_2 in Table 1.

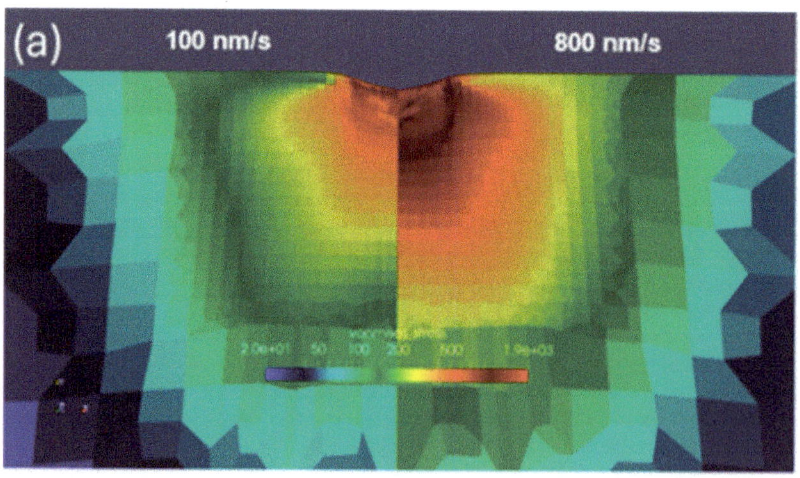

Figure 4. *Cont.*

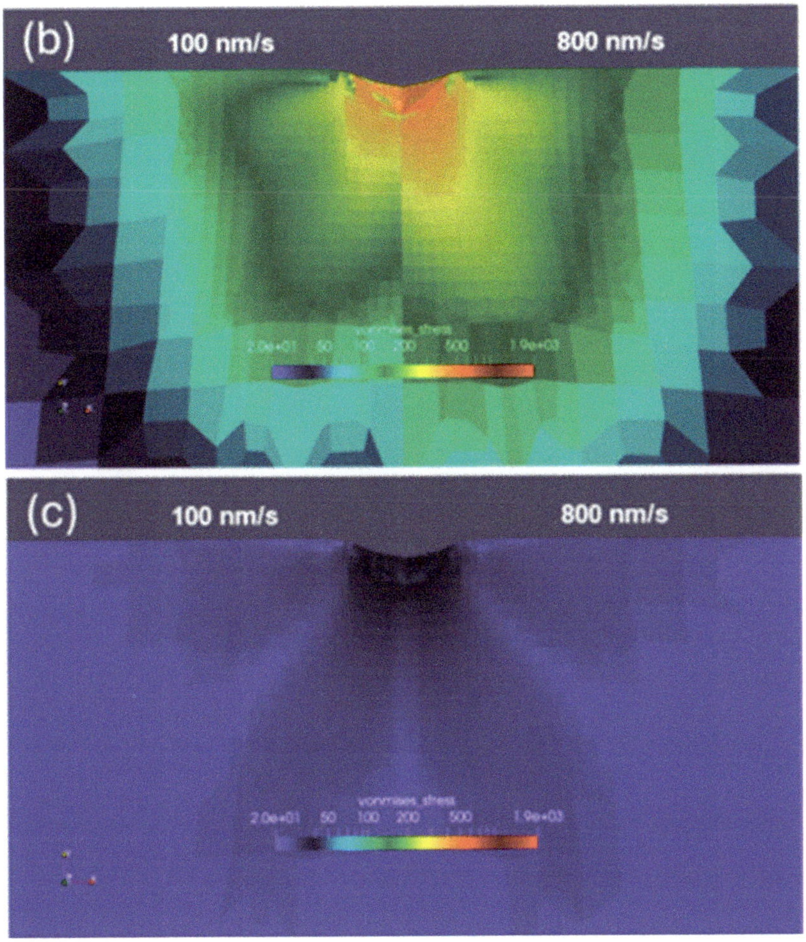

Figure 4. Comparison between von Misses stress distributions of samples pre-loaded at 100 (left) and 800 nm/s (right) after 0, 1, and 15 s following indentation loading in (**a**), (**b**), and (**c**), respectively.

The CP stress relaxation simulation results were further analyzed to show the spatial distribution of the spontaneous dislocation accumulation following different pre-loading rates. In the Kalidindi model [24], the slip resistance, R_S, is employed to characterize the dislocation density, as rationalized by Equation (4). In Figure 5, the changes in R_S within the first 0.002 s of stress relaxation were probed at different depths from the indenter tip, following the pre-loading at 100, 800, and 1600 nm/s. When normalized for comparison, the data show that the dislocation accumulation rate peaks at a shallower position at a greater pre-loading displacement rate. This result, agreeing with the experimental observation in Figure 3, supports our hypothesis that the effective depth of dislocation accumulation during the stress relaxation decreases with an increasing indentation pre-loading rate.

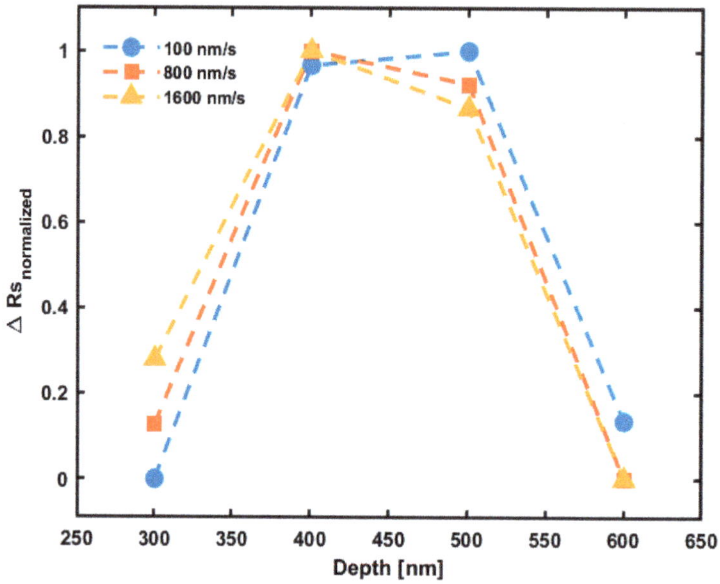

Figure 5. Normalized spontaneous slip resistance changes as a function of depth from indenter tip following indenter pre-loading at 100, 800, and 1600 nm/s loading rates.

4.3. Limitations

The limitations of the data interpretation method for nanoindentation stress relaxation arise from the applicability of the underlying scaling relationships and the constraints of instrumented nanoindentation. The scaling analysis assumes that dislocation mobility is proportional to the network density via the MPF, as described in Equations (3)–(5). However, this relationship does not apply to the dislocation-glide regime, where the high stress enhances the dislocation MPF beyond the limits set by network density.

Moreover, the range of the pre-loading rates for stress relaxation tests is constrained by the instrument's capabilities. If the pre-loading rate is too high, the rapid initial stress relaxation may exceed the system's dynamic response capability. Conversely, if the pre-loading rate is too low, the stress relaxation is so subtle that noise can overshadow the signal.

5. Conclusions

The stress relaxation process under nanoindentation is complicated due to the strong stress and strain gradients. For room-temperature aluminum, at least two mechanisms contribute to stress relaxation, as indicated by the two distinctive characteristic frequencies in Table 2. The more rapid stress relaxation transient, happening right after the indenter is held in position, is attributed to the dislocation accumulation and associated with an increase in internal stress and decrease in effective stress. As a result, the conventional assumptions of constant internal stress or constant dislocation density do not apply to interpreting the nanoindentation stress relaxation data. We verified that the effective depth of dislocation accumulation during the stress relaxation decreases with an increasing indentation pre-loading rate, which is a consequence of the similitude in dislocation kinetics [13]. As a result, the scaling relationship between the stress relaxation kinetics and the indentation pre-loading rate can be utilized to obtain the dislocation velocity–stress exponent, m, from a set of nanoindentation stress relaxation experiments with various pre-loading rates. This new method found $m = 2.5 \pm 0.5$, leading to a strain rate–stress exponent $n = 4.5 \pm 0.5$ for pure aluminum at room temperature. Such consistency with the strain rate–stress exponent determined by the bulk creep experiment has not been achieved by other nanoindentation-based methods and is attributed to the scaling analysis in data interpretation.

Author Contributions: Conceptualization, T.C. and D.M.F.; methodology, T.C., D.Y., S.P. and T.-Y.C.; formal analysis, T.-Y.C. and G.V.; investigation, T.-Y.C. and G.V.; resources, T.C., D.M.F., D.Y. and S.P.; writing—original draft preparation, T.-Y.C. and T.C.; writing—review and editing, T.C., D.M.F., D.Y. and S.P.; visualization, T.-Y.C. and G.V.; supervision, T.C., D.M.F., D.Y. and S.P.; project administration, D.M.F.; funding acquisition, D.M.F., D.Y. and T.C. All authors have read and agreed to the published version of the manuscript.

Funding: Work supported through the Idaho National Laboratory Directed Research & Development (LDRD) Program under Department of Energy Idaho Operations Office Contract DE-AC07-05ID14517.

Data Availability Statement: Data are available upon reasonable request.

Acknowledgments: This research made use of Idaho National Laboratory's High Performance Computing systems located at the Collaborative Computing Center and was supported by the Office of Nuclear Energy of the U.S. Department of Energy and the Nuclear Science User Facilities under Contract No. DE-AC07-05ID14517.

Conflicts of Interest: Author David M. Frazer was employed by the company General Atomics. The remaining authors declare that the research was conducted in the absence of any commercial or financial relationships that could be construed as a potential conflict of interest.

References

1. Herbert, E.G.; Pharr, G.M.; Oliver, W.C.; Lucas, B.N.; Hay, J.L. On the Measurement of Stress–Strain Curves by Spherical Indentation. *Thin Solid Films* **2001**, *5*, 331–335. [CrossRef]
2. Hosemann, P. Small-Scale Mechanical Testing on Nuclear Materials: Bridging the Experimental Length-Scale Gap. *Scr. Mater.* **2018**, *143*, 161–168. [CrossRef]
3. Frazer, D.; Hosemann, P. Plasticity of UO2 Studied and Quantified via Elevated Temperature Micro Compression Testing. *J. Nucl. Mater.* **2019**, *525*, 140–144. [CrossRef]
4. Frazer, D.; Abad, M.D.; Krumwiede, D.; Back, C.A.; Khalifa, H.E.; Deck, C.P.; Hosemann, P. Localized Mechanical Property Assessment of SiC/SiC Composite Materials. *Compos. Part A Appl. Sci. Manuf.* **2015**, *70*, 93–101. [CrossRef]
5. Chen, T.; Frazer, D.; Cullison, M.; Teng, F.; Lian, J.; Mo, K.; Cappia, F. The Effects of Radiation-Induced Grain Subdivision and Dislocations on the Fracture Properties of Uranium Dioxide. *J. Nucl. Mater.* **2023**, *577*, 154297. [CrossRef]
6. Nix, W.D.; Gao, H. Indentation Size Effects in Crystalline Materials: A Law for Strain Gradient Plasticity. *J. Mech. Phys. Solids* **1998**, *46*, 411–425. [CrossRef]
7. Frazer, D.; Shaffer, B.; Gong, B.; Peralta, P.; Lian, J.; Hosemann, P. Elevated Temperature Nanoindentation Creep Study of Plastically Deformed and Spark Plasma Sintered UO2. *J. Nucl. Mater.* **2021**, *545*, 152605. [CrossRef]
8. Chen, T.; Tan, L.; Lu, Z.; Xu, H. The Effect of Grain Orientation on Nanoindentation Behavior of Model Austenitic Alloy Fe-20Cr-25Ni. *Acta Mater.* **2017**, *138*, 83–91. [CrossRef]
9. Chen, T.; He, L.; Cullison, M.H.; Hay, C.; Burns, J.; Wu, Y.; Tan, L. The Correlation between Microstructure and Nanoindentation Property of Neutron-Irradiated Austenitic Alloy D9. *Acta Mater.* **2020**, *195*, 433–445. [CrossRef]
10. Goodall, R.; Clyne, T.W. A Critical Appraisal of the Extraction of Creep Parameters from Nanoindentation Data Obtained at Room Temperature. *Acta Mater.* **2006**, *54*, 5489–5499. [CrossRef]
11. Xu, B.X.; Wang, X.M.; Yue, Z.F. Determination of the Internal Stress and Dislocation Velocity Stress Exponent with Indentation Stress Relaxation Test. *J. Mater. Res.* **2008**, *23*, 2486–2490. [CrossRef]
12. Kutty, T.R.G.; Basak, C.B.; Kumar, A.; Kamath, H.S. Creep Behaviour of δ-Phase of U–Zr System by Impression Creep Technique. *J. Nucl. Mater.* **2011**, *408*, 90–95. [CrossRef]
13. Zaiser, M.; Sandfeld, S. Scaling Properties of Dislocation Simulations in the Similitude Regime. *Model. Simul. Mater. Sci. Eng.* **2014**, *22*, 065012. [CrossRef]
14. Hull, D.; Bacon, D.J. *Introduction of Dislocation*, 5th ed.; Elsevier: Amsterdam, The Netherlands, 2011; ISBN 978-0-7506-4681-9.
15. Blum, W.; Eisenlohr, P.; Breutinger, F. Understanding Creep—A Review. *Metall. Mater. Trans. A* **2002**, *33*, 291–303. [CrossRef]
16. Blum, W. Mechanisms of Creep Deformation in Steel. In *Creep-Resistant Steels*; Elsevier: Amsterdam, The Netherlands, 2008; pp. 365–402. ISBN 978-1-84569-178-3.
17. Varma, A.; Krishnaswamy, H.; Jain, J.; Lee, M.-G.; Barlat, F. Advanced Constitutive Model for Repeated Stress Relaxation Accounting for Transient Mobile Dislocation Density and Internal Stress. *Mech. Mater.* **2019**, *133*, 138–153. [CrossRef]
18. Aifantis, E.C.; Gerberich, W.W. A Theoretical Review of Stress Relaxation Testing. *Mater. Sci. Eng.* **1975**, *21*, 107–113. [CrossRef]
19. Gupta, I.; Li, J.C.M. Stress Relaxation, Internal Stress, and Work Hardening in Some Bcc Metals and Alloys. *Metall. Trans.* **1970**, *1*, 2323–2330. [CrossRef]
20. Dotsenko, V.I. Stress Relaxation in Crystals. *Phys. Status Solidi B* **1979**, *93*, 11–43. [CrossRef]

21. Kruml, T.; Coddet, O.; Martin, J.L. About the Determination of the Thermal and Athermal Stress Components from Stress-Relaxation Experiments. *Acta Mater.* **2008**, *56*, 333–340. [CrossRef]
22. Barlat, F. A Simple Model for Dislocation Behavior, Strain and Strain Rate Hardening Evolution in Deforming Aluminum Alloys. *Int. J. Plast.* **2002**, *18*, 919–939. [CrossRef]
23. Lindsay, A.D.; Gaston, D.R.; Permann, C.J.; Miller, J.M.; Andrš, D.; Slaughter, A.E.; Kong, F.; Hansel, J.; Carlsen, R.W.; Icenhour, C.; et al. 2.0—MOOSE: Enabling Massively Parallel Multiphysics Simulation. *SoftwareX* **2022**, *20*, 101202. [CrossRef]
24. Kalidindi, S.R.; Bronkhorst, C.A.; Anand, L. Crystallographic Texture Evolution in Bulk Deformation Processing of FCC Metals. *J. Mech. Phys. Solids* **1992**, *40*, 537–569. [CrossRef]
25. Karimzadeh, A.; Koloor, S.S.R.; Ayatollahi, M.R.; Bushroa, A.R.; Yahya, M.Y. Assessment of Nano-Indentation Method in Mechanical Characterization of Heterogeneous Nanocomposite Materials Using Experimental and Computational Approaches. *Sci. Rep.* **2019**, *9*, 15763. [CrossRef] [PubMed]
26. Kubát, J.; Rigdahl, M. The Exponential and Power Laws of Stress Relaxation Kinetics and a General Relation between the Activation Volume and Effective Stress. *Phys. Status Solidi A* **1976**, *35*, 173–180. [CrossRef]
27. Hérault, Q.; Gozhyk, I.; Balestrieri, M.; Montigaud, H.; Grachev, S.; Lazzari, R. Kinetics and Mechanisms of Stress Relaxation in Sputtered Silver Thin Films. *Acta Mater.* **2021**, *221*, 117385. [CrossRef]
28. Caillard, D.; Martin, J.-L. *Thermally Activated Mechanisms in Crystal Plasticity*; Pergamon: Amsterdam, The Netherlands, 2003; ISBN 978-0-08-042703-4.
29. Zinkle, S.J. Radiation-Induced Effects on Microstructure. In *Comprehensive Nuclear Materials*; Elsevier: Amsterdam, The Netherlands, 2020; pp. 91–129. ISBN 978-0-08-102866-7.
30. Kassner, M.E.; Pérez-Prado, M.-T. *Fundamentals of Creep in Metals and Alloys*, 1st ed.; Elsevier: Amsterdam, The Netherlands; Heidelberg, Germany, 2004; ISBN 978-0-08-043637-1.
31. Ueda, S.; Kameyama, T.; Matsunaga, T.; Kitazono, K.; Sato, E. Re-Examination of Creep Behaviour of High Purity Aluminium at Low Temperature. *J. Phys. Conf. Ser.* **2010**, *240*, 012073. [CrossRef]
32. Gibbs, G.B. Creep and Stress Relaxation Studies with Polycrystalline Magnesium. *Philos. Mag. J. Theor. Exp. Appl. Phys.* **1966**, *13*, 317–329. [CrossRef]
33. Kombaiah, B.; Murty, K.L. Dislocation Cross-Slip Controlled Creep in Zircaloy-4 at High Stresses. *Mater. Sci. Eng. A* **2015**, *623*, 114–123. [CrossRef]

crystals MDPI

Communication

Improving Texture Prediction by Increasing Mesh Resolution in Submodel: A Crystal Plasticity FE Study and Experiment Verification

Yu Liu [1,2], Qi Zhang [2], Qinqin Ge [3], Xingxing Wang [2] and Yifu Shen [1,*]

1 College of Materials Science and Technology, NUAA, Nanjing 211106, China; liuyu_me@ntu.edu.cn
2 School of Mechanical Engineering, Nantong University, Nantong 226019, China; 2110110128@stmail.ntu.edu.cn (Q.Z.); wangxx@ntu.edu.cn (X.W.)
3 Mechanical Maintenance Department, Nantong Cellulose Fibers Co., Ltd., Nantong 226007, China; 11317@ncfcinfo.com
* Correspondence: yfshen_nuaa@hotmail.com

Abstract: Crystal plasticity finite element simulations require tremendous computation time and, accordingly, coarse mesh is generally used. To improve the texture prediction, Submodelling was applied to feature grains in this study. A simulation of the Wholemodel (whole sample) was firstly carried out to obtain the global texture, and then a smaller region from the Wholemodel was selected, reconstructed and finely meshed in the Submodel. The movement on the selected region boundary, obtained from the Wholemodel, was used to deform the Submodel. The Submodel reproduced the predictions in the Wholemodel, and the texture prediction, especially at micro-scale, was greatly enhanced in the Submodel due to the fine mesh. This significant drop in the Submodel computation time marks an ~85% decrease compared to the Wholemodel.

Keywords: crystal plasticity FE; Submodel; mesh resolution; texture

1. Introduction

Texture modelling has become a powerful tool to predict metal deformation behaviours from macro- to micro-scale (e.g., [1,2]). Among these various crystal plasticity (CP) models, the so-called 'full-field' CP finite element (CPFE) model is supposed to be the best one for modelling the texture evolution [2–4]. The CPFE model introduces CP constitutive laws into the finite element (FE) package; the FE solves the boundary problems and the CP model is used to determine the plastic deformation at material points [5]. In each FE increment, the equilibrium in all elements is met with fundamental mechanical principles [2], which results in an obvious demerit: huge computation time. Increasing the mesh resolution is beneficial to improving the texture prediction [4,6–8], but reducing the element size results in a huge computation time. This explains why the mesh refinement ratios are very low in the literature, usually below 100 [5,6]. In this research, Submodelling was applied to a small region with fine mesh. The Submodel successfully reproduced the predictions in the Wholemodel, and the computation time was exceedingly reduced.

2. Materials and Experiments

High-purity aluminium alloy sheets were processed using rolling in this work. The material was fully annealed at 450 °C for 2 h, and sheet materials were cut into a dimension of 200 mm (length) × 40 mm (width) × 2 mm (thickness). The rolling at 50% reduction was performed on a rolling mill, consisting of 125 mm diameter rolls. After rolling, electron backscattered diffraction (EBSD) scans were performed to characterize the through-thickness textures on the rolling direction (RD)–normal direction (ND) plane. The EBSD samples were taken from the sheet centre, and the cross-section was mechanically ground and then polished up to 1 μm and finally electro-polished (16 °C, 25 V) in an A3 Struers electrolyte. EBSD scans were performed with a working distance of 15 mm and an accelerating voltage of 15 kV. HKL Channel-5 was used to postprocess the EBSD scanning.

3. Submodel Modelling

When employing FE modelling, there is a trade-off between computation time and mesh resolution, and thus the mesh refinement ratios are usually low. Submodelling is an alternative to assess a feature region, as this method is schematically shown in Figure 1. A coarse-mesh Wholemodel (whole sample) is firstly developed and then numerically solved. A critical part is selected and reconstructed in a fine-mesh Submodel. The displacement of nodes on the driven surface (in blue), which is extracted from the Wholemodel, acts as the boundary conditions for the Submodel, while the inner nodes (in green) are free to move according to the Submodel deformation.

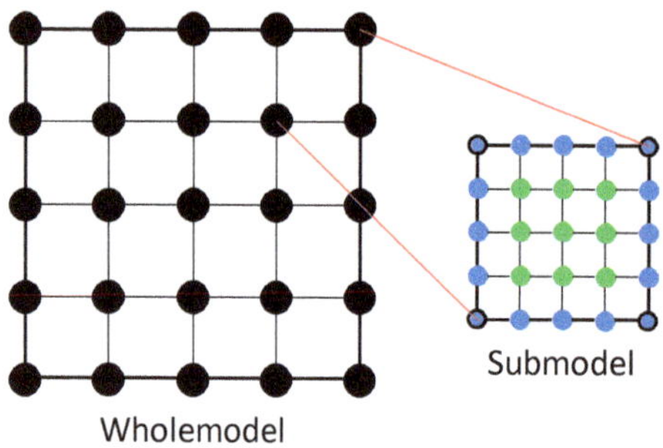

Figure 1. Submodel modelling method.

The simulation, following the experimental conditions and procedure, was carried out using the commercial FE package Abaqus/Standard ver.6.9. A two-dimensional FE model under plain-strain conditions was developed. The rolls (125 mm in diameter) were modelled as analytical rigid bodies. The element type is CPE4R, four-node elements with a reduced integration point. A polycrystal model with a coarse mesh (element size: 50 μm × 50 μm), i.e., Wholemodel, was developed, as shown in Figure 2. After the Wholemodel, a feature region, Grain A in Figure 2, was reconstructed in the Submodel-1 and was then finely meshed with refinement ratios of 1:25 and 1:625. In the former and latter cases, one Wholemodel element was divided into 25 elements (element size: 10 μm × 10 μm) and 625 elements (element size: 2 μm × 2 μm), respectively. The element size of 2 μm × 2 μm in the Submodel-1 is comparably large as the step size used in EBSD scans. In the Submodel-2 (Figure 2), not only Grain A but also its neighbouring grains were included. Grain A in the Submodel-2 was driven by the neighbouring grains, which are supposed to apply relatively 'soft' boundary conditions on Grain A. When Submodelling, the initial grain orientations in both the Wholemodel and Submodel were kept the same.

The well-recognized kinematical scheme developed by Asaro [8] and the widely adopted Bassani–Wu hardening model [9,10] were used in this study. The material parameters were identified by matching the predictions to the experimental results compression [11]. Textures in different processes have been successfully captured by this CPFE model [3,12,13], which proves the reliability and capability of this CPFE model. The crystal plasticity constitutive law, hardening model and material parameters, as well as the CPFE implement, have been given in Ref. [3].

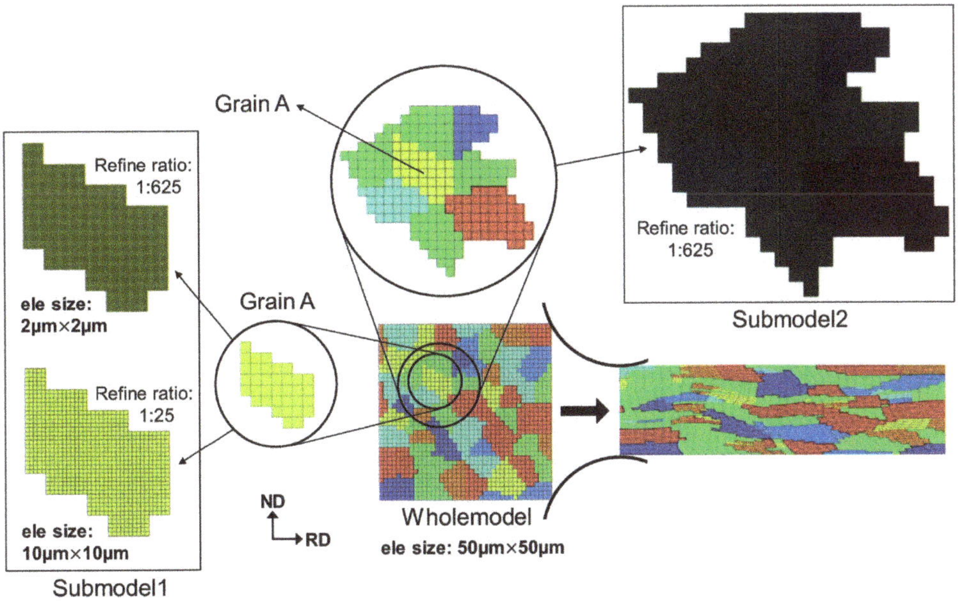

Figure 2. Mesh resolution of selected grains in Submodel and Wholemodel.

4. Results

Figure 3a,b show a (1 1 1) pole figure and EBSD map of the starting material, respectively. The initial texture is a recrystallized texture, i.e., [5] <0 0 1> Cube dominated (Figure 3a), and only a small number of low-angle grain boundaries (LAGBs) ($2° \leq \theta < 15°$) in grains can be seen (Figure 3b). After rolling, the global texture evolved into a rolling texture (Figure 3c), typically consisting of {1 1 2}<1 1 1>, {1 2 3}<6 3 4> and {1 1 0}<1 1 2>. The high-angle grain boundaries (HAGBs) in Figure 3d are believed to be from the original ones but did not evolve during rolling, and the subgrain and LAGBs formed within the initially annealed grains. The pole figures of three representative grains are shown on the right of Figure 3d, from which it can be seen that an in-grain subdivision occurred and initially uniform in-grain orientations became scattered.

The predicted global texture from the Wholemodel (Figure 3e) reproduces the experimental observation in Figure 3c. After rolling, texture gradients developed in the Wholemodel (Figure 3f), which means an in-grain subdivision also occurred. However, compared to the step size in Figure 3d, the mesh resolution in the Wholemodel (Figure 3f) was exceedingly low. From Figure 3, it can be concluded that an acceptable global texture can be captured by the coarse-mesh Wholemodel.

The texture evolution of Grain A in the Wholemodel is shown in Figure 4a, and only a few elements were used to describe the in-grain texture; thus, the texture gradient is very high. As for the Submodel-1, a similar texture of Grain A was reproduced with the refinement ratio of 1:25 (Figure 4b), where the intensities of major texture components were strengthened and subtle components were also captured. The texture prediction was further improved by increasing the refinement ratio to 1:625 (Figure 4c), and macroscopic shear bands (marked by black arrows) and microscopic shear bands (red arrows) were also predicted. The predictions in the Submodel-2 1:625 (Figure 4d) are similar to those in Figure 4c in terms of shear bands and strong and weak texture components, although the grain shape is slightly different, since Grain A in the Submodel-2 was driven by the neighbouring grains, but not directly by its deformation history in the Wholemodel. The crystal rotation angles in the four models (shown in Figure 4) are similarly distributed in Figure 5. Although the fraction of relatively large rotation angles decreases slightly as the mesh resolution increases, the distributions of all four models are generally consistent. This agrees with the slightly lowered average rotation angles of Grain A in the four models (Figure 4).

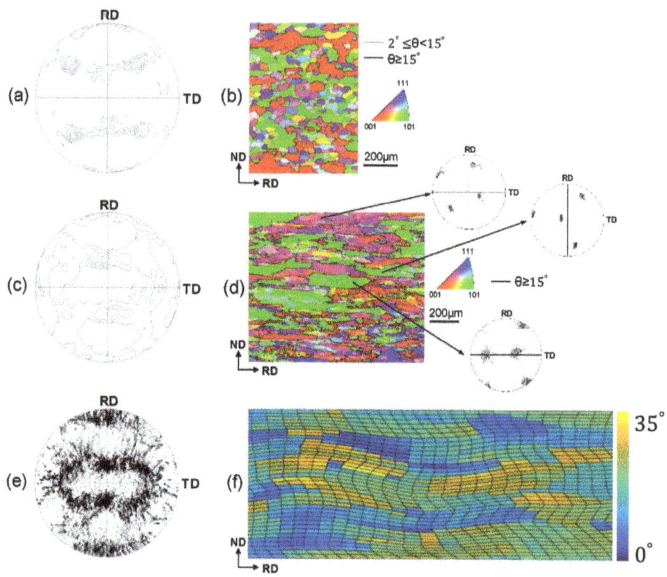

Figure 3. (1 1 1) pole figures of the global texture (**a**) before rolling, (**c**) after rolling and (**e**) in the Wholemodel. IPF EBSD maps (**b**) before rolling and (**d**) after rolling, and (**f**) distribution of crystal rotation angles in the Wholemodel.

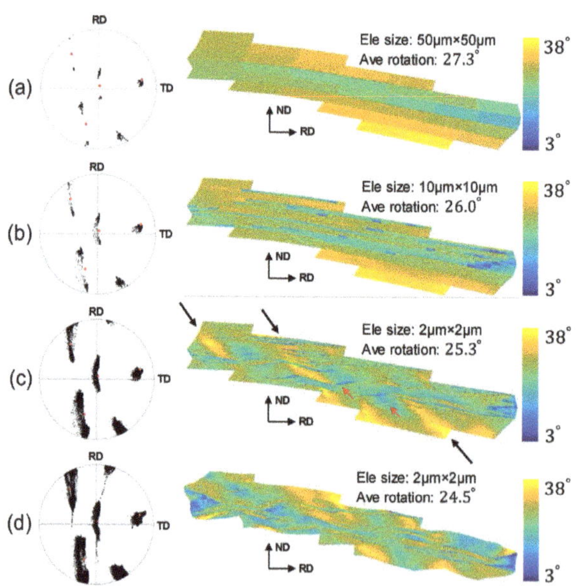

Figure 4. (1 1 1) pole figures and distribution of crystal rotation angles of Grain A in the (**a**) Wholemodel, (**b**) Submodel-1 1:25, (**c**) Submodel-1 1:625 and (**d**) Submodel-2 1:625. The initial orientations are shown in red, and final orientations in black. Black arrows indicate grain-scale shear bands, while micro-scale shear bands are marked by red arrows.

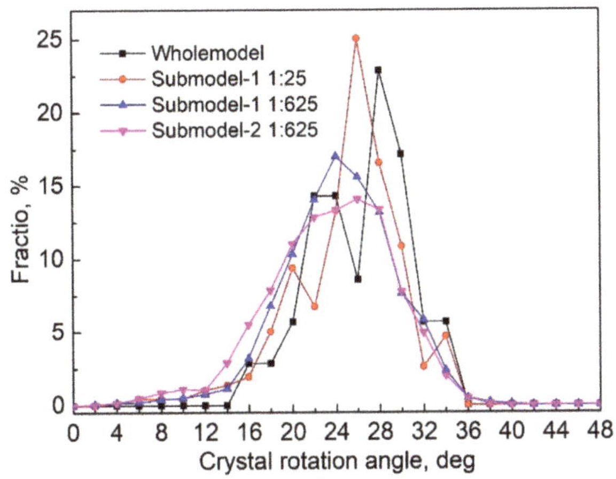

Figure 5. Distribution of crystal rotation angles in Grain A.

Figure 6 shows the CPU time at different refinement ratios. It can be seen that the computation time of the Wholemodel was reduced by about ~85% in the Submodel when the element number in them is the same. This is because complicated boundary conditions and interactions were considered when analysing the Wholemodel, and they cost expensive computation sources, while the boundary conditions for the Submodel are greatly simplified and thus the computation time is reduced. As for the Submodel with different mesh refinement ratios, it can be seen that the CPU time is almost linearly related to the element number. The CPU time increases ~17 times when the mesh refinement ratio is 625. When using the Submodel method, the computation time became affordable, especially for a relatively small region.

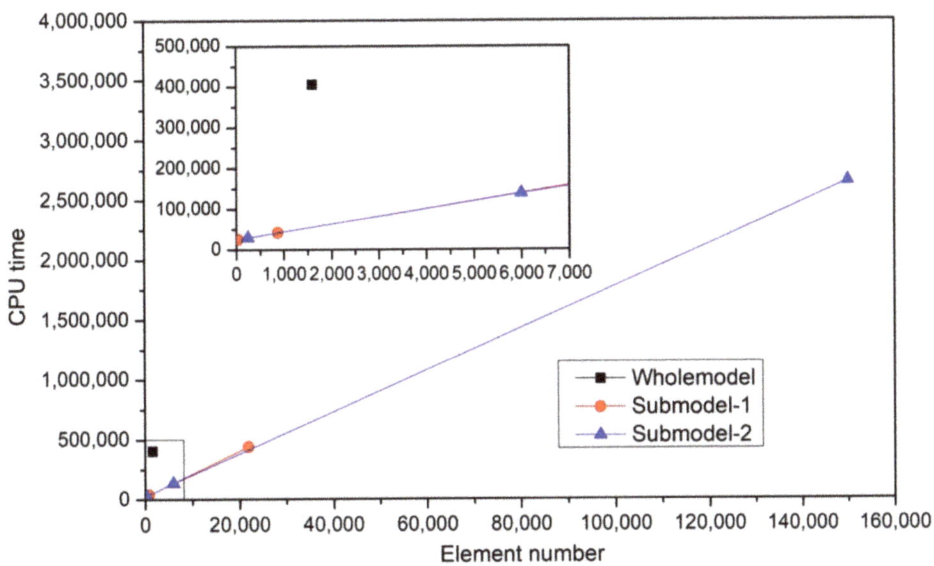

Figure 6. Relation between element number and CPU time in seconds.

5. Discussion

No homogenization is theoretically assumed in the CPFE model, but intra-element subdivision is not accessible (saying element type is CPE4R). Though the element size in the Wholemodel was obviously larger than the EBSD step sizes, acceptable global textures have been predicted by the coarse mesh. In contrast, the texture prediction was greatly enhanced by reducing the element size in the Submodel in terms of major and minor texture components, as well as macro- and micro-scale shear bands. Increasing the mesh resolution means more elements are used to describe the textural gradients, so the subtle textures can be reasonably captured. Meanwhile, the material flow was enhanced, and this is why in-grain shear bands (Figure 4c,d) were predicted. Inherent length scale is not assumed in the current CPFE model [5,6] and there is no assumed homogenization, which implies that similar textures to those in the Wholemodel [5] can be reproduced in the Submodel. The boundary conditions of the Submodel were slightly altered (Figure 4d), but the alteration is negligible, as can be seen from the comparison of grain shapes and averaged rotation angles between the four models (Figures 4 and 5). Moreover, compared to external loading, the initial orientations of grains played a dominant role in the in-grain texture development [12]. With the affordable computation time, the refinement ratio reached a high number due to the reduced size of the sample in the Submodel. The maximum mesh refinement ratio is 625 in this study, but it is only 16 [5] and 30 [6] when fine meshes were applied to the whole sample. As an alternative, Submodelling is capable of reproducing predictions at an exceedingly reduced computation time. This study proves that Submodelling can be applied to feature regions in various FE simulations.

6. Conclusions

1. The global texture produced by the coarse-mesh Wholemodel agrees well with the experimental observations.

2. Submodelling was adopted to access the texture in a grain with highly refined mesh. The Submodel successfully reproduced the predictions in the Wholemodel, and more experimentally observed phenomena such as subtle texture components and shear bands were revealed in the Submodel.

3. The computation time for the Submodel was reduced by ~85% compared with the Wholemodel, because the complicated boundary conditions and interactions in the Wholemodel, which need much computation sources, were not considered in the Submodel.

4. Compared to the Wholemodel, the computation time of the Submodel was greatly saved due to the reduced region, and thus high-mesh resolution can be used in the Submodel. This study proves that Submodelling can be applied to feature regions in various FE simulations.

Author Contributions: Conceptualization, Y.L..; Data curation, Q.Z.; Formal analysis, Q.G. and Y.L.; Investigation, X.W.; Methodology, Y.L. and Y.S.; Project administration, Y.S.; Resources, Y.S.; Software, Y.L.; Supervision, Y.S.; Validation, Q.Z.; Visualization, Q.G.; Writing—original draft, Y.L.; Writing—review & editing, Y.L. All authors have read and agreed to the published version of the manuscript.

Funding: This research was funded by Natural Science Horizontal Research Project of Nantong University (grant number: 22ZH643) and A Project Funded by the Priority Academic Program Development of Jiangsu Higher Education Institutions (PAPD).

Data Availability Statement: Not applicable.

Acknowledgments: The authors would like to express their gratitude for the Natural Science Horizontal Research Project of Nantong University (grant number: 22ZH643) and A Project Funded by the Priority Academic Program Development of Jiangsu Higher Education Institutions (PAPD).

Conflicts of Interest: The authors declare no conflict of interest.

References

1. Zhang, K.; Holmedal, B.; Hopperstad, O.S.; Dumoulin, S.; Gawad, J.; Van Bael, A.; Van Houtte, P. Multi-level modelling of mechanical anisotropy of commercial pure aluminium plate: Crystal plasticity models, advanced yield functions and parameter identification. *Int. J. Plast.* **2015**, *66*, 3–30. [CrossRef]
2. Van Houtte, P.; Li, S.; Seefeldt, M.; Delannay, L. Deformation texture prediction: From the Taylor model to the advanced Lamel model. *Int. J. Plast.* **2005**, *21*, 589–624. [CrossRef]

3. Wang, H.; Cheng, L.; Kiet, T.; Yu, L. A crystal plasticity FE study of macro- and micro-subdivision in aluminium single crystals {001}<110> multi-pass rolled to a high reduction. *J. Mater. Sci. Technol.* **2021**, *76*, 231–246.
4. Zhao, Z.; Kuchnicki, S.; Radovitzky, R.; Cuitiño, A. Influence of in-grain mesh resolution on the prediction of deformation textures in fcc polycrystals by crystal plasticity FEM. *Acta Mater.* **2007**, *55*, 2361–2373. [CrossRef]
5. Buchheit, T.E.; Wellman, G.W.; Battaile, C.C. Investigating the limits of polycrystal plasticity modelling. *Int. J. Plast.* **2005**, *21*, 221–249. [CrossRef]
6. Harewood, F.J.; McHugh, P.E. Investigation of finite element mesh independence in rate dependent materials. *Comput. Mater. Sci.* **2006**, *37*, 442–453. [CrossRef]
7. Inoue, T.; Tsuji, N. Quantification of strain in accumulative roll-bonding under unlubricated condition by finite element analysis. *Comput. Mater. Sci.* **2009**, *46*, 261–266. [CrossRef]
8. Asaro, R.J. Crystal Plasticity. *J. Appl. Mech. Trans. ASME* **1983**, *50*, 921–934. [CrossRef]
9. Wu, T.-Y.; Bassani, J.L.; Laird, C. Latent Hardening in Single Crystals I. *Theory Exp. Proc. R. Soc. A Math. Phys. Eng. Sci.* **1991**, *435*, 20.
10. Bassani, J.L.; Wu, T.-Y. Latent Hardening in Single Crystals II. *Anal. Charact. Predict. Proc. R. Soc. A Math. Phys. Eng. Sci.* **1991**, *435*, 21.
11. Wang, H.; Cheng, L.; Kiet, T. A Combined Experiment and Crystal Plasticity FEM Study of Microstructure and Texture in Aluminium Processed by Reverse and Unidirectional Accumulative Roll-Bonding. *Crystals* **2019**, *9*, 119. [CrossRef]
12. Wei, P.; Lu, C.; Tieu, K.; Su, L.; Deng, G.; Huang, W. A study on the texture evolution mechanism of nickel single crystal deformed by high pressure torsion. *Mater. Sci. Eng. A* **2017**, *684*, 239–248. [CrossRef]
13. Raabe, D.; Zhao, Z.; Mao, W. On the dependence of in-grain subdivision and deformation texture of aluminum on grain interaction. *Acta Mater.* **2002**, *50*, 4379–4394. [CrossRef]

MDPI AG
Grosspeteranlage 5
4052 Basel
Switzerland
Tel.: +41 61 683 77 34

Crystals Editorial Office
E-mail: crystals@mdpi.com
www.mdpi.com/journal/crystals

www.ingramcontent.com/pod-product-compliance
Lightning Source LLC
Chambersburg PA
CBHW062050220926
51850CB00010B/2724
* 9 7 8 3 7 2 5 8 5 7 1 1 1 *